Algorithms and
Data Structures
in VLSI Design

T0238304

Springer
Berlin
Heidelberg
New York
Barcelona
Budapest
Hong Kong
London
Milan
Paris
Singapore
Tokyo

Christoph Meinel
Thorsten Theobald

Algorithms and Data Structures in VLSI Design

OBDD – Foundations and Applications

With 116 Figures

 Springer

Prof. Dr. Christoph Meinel

University of Trier
D-54286 Trier, Germany

Dr. Thorsten Theobald

Technical University München
D-80290 München, Germany

Title of the Original German Edition:
Algorithmen und Datenstrukturen im VLSI-Design
OBDD – Grundlagen und Anwendungen
© Springer-Verlag Berlin Heidelberg 1998

Library of Congress Cataloging-in-Publication Data

Meinel, Christoph, 1954–
  [Algorithmen und Datenstrukturen im VLSI-design. English]
  Algorithms and data structures in VLSI design: OBDD-foundations
and applications/Christoph Meinel, Thorsten Theobald.
    p. cm.
  Includes bibliographical references and index.
  ISBN 3-540-64486-5 (softcover: alk. paper)
  1. Integrated circuits – Very large scale integration – Design and
construction – Data processing. 2. Computer algorithms. 3. Data
structures (Computer science) I. Theobald, Thorsten, 1971–   .
II. Title.
  TK7874.75.M45 1998
  621.39'5–dc21   98–26197
                                                              CIP

ISBN 3-540-64486-5 Springer-Verlag Berlin Heidelberg New York

© Springer-Verlag Berlin Heidelberg 1998
Printed in Germany

The use of general descriptive names, trademarks, etc. in this publication does not imply, even in
the absence of a specific statement, that such names are exempt from the relevant protective laws
and regulations and therefore free for general use.

Typesetting: Camera-ready by the authors
Cover Design: Künkel + Lopka, Heidelberg
Printed on acid-free paper   SPIN 10679136   33/3142 – 5 4 3 2 1 0

Preface

Within the last few years, an exciting interaction has taken place between current research in theoretical computer science and practical applications in the area of VLSI design. This has mainly been initiated by the following two facts. Firstly, chip design without the assistance of computers is no longer conceivable. And secondly, the construction of permanently more powerful chips is reaching the absolute frontiers of current possibilities and capabilities. In 1994, the design error in the Intel Pentium chip with its devastating economical consequences has made particularly aware of these difficulties.

One of the main problems is the huge number of possible combinations of individual chip elements (giving a *combinatorial explosion*). For such chips, the problem of representing and manipulating the functional behavior of the chip within a computer has already become almost impossible. Here, a fruitful connection with a core topic of theoretical computer science, the *design of data structures and efficient algorithms*, has been achieved. This connection is established by means of *ordered binary decision diagrams (OBDDs)*, which have led to dramatic performance improvements and breakthroughs in many CAD projects all over the world.

This book provides a comprehensive introduction to the foundations of this interdisciplinary research topic and discusses important applications in computer-aided circuit design. It originated from undergraduate and graduate courses at the University of Trier as well as from relevant research projects.

On the one hand, the book is directed to students, researchers and lecturers who work or teach in the areas of algorithms, data structures, or VLSI design and are looking for access to the active research field of OBDDs, or are interested in paradigmatic connections between theory and practice. On the other hand, the book is intended to provide a valuable source for all interested CAD engineers, VLSI designers, and users of CAD tools who are looking for a self-contained presentation of OBDD technology, or are interested in questions of modeling and methodology in CAD projects.

The general prerequisites for reading the book are only some basic knowledge in computer science and pleasure in studying clever and ingenious ideas that make CAD tools work efficiently.

At this point, we would like to thank everybody who was directly or indirectly involved in the process of writing this book. In particular, we would like to thank the German Research Foundation (Deutsche Forschungsgemeinschaft, DFG) which supported our research activities within the major program "Efficient algorithms for discrete problems" and within the graduate program "Mathematical optimization".

Of course, our collaborators and colleagues Jochen Bern, Jordan Gergov, Stefan Krischer, Harald Sack, Klaus Schwettmann, Anna Slobodová, and Christian Stangier have influenced the work substantially. Further thanks are also extended to Prof. Wolfram Büttner and his working group at Siemens for integrating us into current OBDD-based research projects there. With regard to our numerous foreign colleagues, we would particularly like to emphasize the help and support of Fabio Somenzi, who was an invaluable source of information, for example during a research visit lasting several weeks in Trier. Finally, we would like to thank Peter Gritzmann, whose willingness and support allowed us to finish this book project.

Trier – München   *Christoph Meinel*

June, 1998   *Thorsten Theobald*

Contents

Part III. Applications and Extensions

1. Introduction

A wide variety of problems in discrete mathematics, computer science, or the design of integrated circuits involve the task of manipulating objects over finite domains. The dependencies and relations among the objects are modeled by means of discrete functions. By introducing a binary encoding for the elements of the underlying domain, each finite problem can be fully expressed by means of *switching functions*, i.e., functions mapping bit vectors to single bits. Consequently, manipulations over finite domains can be entirely reduced to the treatment of switching functions.

This approach is of particular importance in *computer-aided design (CAD)* of digital systems. In this area, recent advances in *very large scale integration (VLSI)* technology have raised problems that are far beyond the scope of manual design. VLSI design in the absence of computer-aided design systems is no longer imaginable. Hence, CAD systems have become not only a useful tool but also an indispensable component of every chip design process.

However, the real capabilities and power of those CAD systems strongly depend on the following two factors:

1. *Compactness* of the *data structures* used for representing relevant data and switching functions within the computer.
2. *Efficiency* of the *algorithms* that operate on the data structures.

Within the last few years, *binary decision diagrams* have attracted much attention in this context. These graphs are composed from a set of binary-valued decisions, culminating in an overall decision that can be either TRUE or FALSE. Binary decision diagrams, shortly called *BDDs*, were proposed as data structure for switching functions by Lee as early as 1959, and later by Akers. Meanwhile, many variations of this basic model have been investigated. In the original model due to Akers, each decision is based on the evaluation of an input variable. Subsequent approaches also consider more highly structured BDDs, or BDDs whose decisions are based on the evaluation of certain functions. From the complexity theoretical point of view, BDDs have been investigated extensively as a computation model related to memory consumption.

In 1986, Bryant showed that typical tasks for manipulating switching functions can be performed very efficiently in terms of Akers' model, if some additional ordering restrictions concerning the structure of the BDD are satisfied. For this reason, Bryant proposed to employ so-called *ordered binary decision diagrams* (*OBDDs*), where each variable is evaluated at most once on each path (*read-once property*), and where the order in which the variables are evaluated does not depend on the input.

OBDDs have found more practical applications than other representations of switching functions, mainly for two reasons. First, by applying reduction algorithms, OBDDs can be transformed into a *canonical form*, called *reduced OBDDs*, which uniquely characterize a given function. Second, in terms of their OBDD representation, Boolean operations on switching functions can be performed quite efficiently in time and space. For example, an AND composition of two switching functions can be performed in time that is linear in the product of the sizes of their OBDD representations.

However, OBDDs share a fatal property with all kinds of representations of switching functions: the representations of *almost all* functions need exponential space. This property is a consequence of an elementary counting argument which was first applied by Shannon. There is another problem that occurs when representing switching functions by means of OBDDs: the size of OBDDs depends on the *order* of the input variables. For example, OBDDs that represent adder functions are very sensitive to the variable order. In the best case, they have linear size in the number of input bits, but in the worst case they have exponential size. Other functions like multiplication functions have OBDDs of exponential size regardless of the variable order. Fortunately, for most functions that appear in real-life applications, a variable order can be found that keeps the size of the corresponding OBDD tractable. Hence, for most practical applications, OBDDs provide a computationally efficient device for manipulating switching functions.

Nowadays, there exist many improvements and additions to the basic OBDD model. Here, we primarily refer to implementation techniques based on *dynamic data structures*. Indeed, using these data structures in connection with certain sophisticated hash techniques allows an efficient realization of OBDD technology in the form of so-called *OBDD packages*. The big success of OBDDs in numerous practical applications is due to these packages. Besides employing the mentioned implementation ideas, structural extensions to the OBDD model have been proposed in order to make practical work as efficient as possible. As an example, we mention complemented edges, which allow the complement of a function to be stored with almost no extra costs.

Meanwhile, for a wide spectrum of purposes, many software systems have been developed that use OBDD packages internally for manipulating the relevant switching functions. The first universal OBDD package was designed by Brace, Rudell, and Bryant. This package, whose design also forms the basis of more recent OBDD packages, exploits the above mentioned ideas in a skillful way. For example, it uses a *computed table* for remembering the results of previous computations, and it supports an ingenious *memory management*.

An important application of OBDDs can be found in *formal verification* of digital circuits. Here, for example, it is to check whether two given combinational circuits have the same logical behavior. In the case of combinational circuits this task can be performed by means of OBDDs as follows. Initially, a suitable variable order is determined, and then the uniquely determined OBDDs of the circuits with respect to the chosen order are constructed. As logical operations on OBDDs can be performed quickly, this computation can be done in the following way. First, the trivial OBDD representations of the input variables are constructed. Then, traversing the circuit in topological order, in each gate the corresponding logical operation is performed on the OBDD representations of the predecessor gates. Due to canonicity of reduced OBDDs, the circuits are logically equivalent if and only if the computed OBDD representations of both circuits are equal.

Other applications of OBDDs are based on the fact that switching functions can be used to represent sets of bit vectors. As an example, we consider a distinguished subset of bit vectors within the set of all n-dimensional bit vectors. The characteristic function of this distinguished subset is a switching function and can therefore be manipulated by means of OBDD techniques. This connection allows to solve problems in *sequential analysis and verification*, where compact representation of state sets is of crucial importance.

The practicability of OBDDs strongly depends on the existence of suitable algorithms and tools for minimizing the graphs in the relevant applications. Within the last few years, primarily two kinds of technique have been investigated, both based on the optimization of the *variable order*. One kind includes heuristic methods, which deduce a priori some information from the application about a good order, and the other involves algorithms for *dynamic reordering*. In many cases, both techniques lead to drastic reductions of the OBDD sizes occurring and hence improve the performance of the CAD system considerably.

Although OBDDs now provide an efficient data structure for solving CAD problem instances far beyond the scope of previous technologies, new frontiers of intractability are coming into view. One critical aspect in the use of OBDDs is their big memory consumption in case of certain complex functions. Hence, the question arises of whether there are more sophisticated BDD-based representations which are more succinct and space-efficient than OBDDs, yet possess similar nice algorithmic properties. Indeed, the improvement of BDD-based models is currently an active research area. So far, these efforts have led to a variety of related data structures with remarkable properties. These data structures include *free BDDs, ordered functional decision diagrams, zero-suppressed BDDs, multi-terminal BDDs, edge-valued BDDs*, and *binary moment diagrams*.

As the practical applicability of all these representations cannot be finally judged, the investigation and development of new optimization techniques for OBDDs remains a rewarding research project. One recent approach proposes to transform switching functions in a suitable way before manipulating them. This approach makes it possible to manipulate functions with inherently large OBDD representations, as long as there exists at least a tractable OBDD representation of a transformed version of the function.

•

At this point, we would like to end the above general survey on some of the topics treated in the text. Let us continue by giving an outline of the book.

In Chapter 2, we review the relevant basics from computer science and discrete mathematics, in particular from the areas of graph theory, algorithms, data structures, and automata theory. The following chapters are organized in three parts.

Part I: Data Structures for Switching Functions

This part deals with basic concepts for representing and manipulating switching functions. In Chapter 3, we begin at the roots with Boolean algebra and Boolean functions. The second half of the chapter is devoted to the special case of switching algebra and switching functions.

Chapter 4 can be seen as a presentation of the world of VLSI design in the period of time before the invention of OBDDs. We discuss classical data structures for switching functions, such as disjunctive normal forms, Boolean formulas, and branching programs. We also emphasize the drawbacks of the particular data structures, which manifest themselves in difficulties in handling them algorithmically.

At the end of Part I, in Chapter 5, important paradigmatic applications in VLSI design are extracted, which allows to deduce the basic requirements on favorable data structures in VLSI design.

Part II: OBDDs: An Efficient Data Structure

Throughout the second part, ordered binary decision diagrams, shortly called OBDDs, are at the center of attention. In Chapter 6, after presenting some introductory examples, we start by discussing the reduction rules. By using these techniques, the two fundamental properties of OBDDs are deduced, namely canonicity in representation and efficient application of binary operations.

Chapter 7 is devoted to implementation techniques that transform the basic concept of OBDDs into an efficient data structure for practical applications.

The first half of the chapter serves to introduce the most important implementation ideas, like the unique table, the computed table, and complemented edges. Then we describe some well-known OBDD software packages in which these ideas have been realized.

In Chapter 8, the influence of the variable order on the complexity of OBDDs is at the center of interest. By applying methods from theoretical computer science, important complexity theoretical properties are proven. In particular, functions which provably do not possess good OBDDs are constructed, complexity problems in the treatment of OBDDs with different orders are discussed, and it is shown that determining the optimal variable order is an **NP**-hard problem.

In Chapter 9, we address practical issues of the ordering problem. We discuss useful heuristics for constructing good orders as well as the powerful technique of dynamic reordering. At the end of the chapter, by means of some benchmark circuits, we demonstrate how these methods prove their worth in practical environments.

Part III: Applications and Extensions

The third part of the book is devoted to various applications as well as to extensions of the OBDD data structure.

The first extensively treated application refers to verification problems of sequential systems, like the equivalence problem. By applying suitable algorithms to concrete instances of these problems, it can be formally proven that the input/output behavior of a given sequential circuit coincides with the behavior of a given reference circuit. The OBDD-based approach, presented in Chapter 10, is based on reachability analysis. In order to realize this strategy efficiently, we first introduce some more complex operators on OBDDs, and then show that the basic step of sequential analysis, namely image computation, can be performed efficiently.

A deeper treatment of this application in the area of model checking can be found in Chapter 11. Here, the task is to check whether a circuit satisfies a specification given by means of logical formulas. First, we give an introduction to the temporal logic CTL, then we use this logic for solving the verification task. This chapter, too, is closed by the description of some real systems in which the presented ideas have been implemented.

Chapter 12 deals with some variants and extensions of the OBDD model. We demonstrate the advantages of those variants and illustrate applications where a certain variant is particularly suited. The chapter starts with a description of free BDDs, which result from OBDDs by weakening the ordering restrictions. Then we discuss functional decision diagrams, which are based on alternative decomposition types. Zero-suppressed BDDs, the variant presented next, are particularly suited for solving combinatorial problems.

We close the chapter by discussing some variants for the representation of multiple-valued functions.

In Chapter 13, we discuss a very general approach to the optimization of decision diagrams which is based on applying suitable transformations to the represented functions. Two particular classes of those transformations, namely type-based transformations and linear transformations, are investigated in more detail. Moreover, in conjunction with analysis of finite state machines, encoding transformations can be applied for optimizing OBDDs. Some aspects of this approach are presented at the end of the chapter.

2. Basics

2.1 Propositions and Predicates

Propositional logic. The theme of **propositional calculus** is to investigate simple logical connections among elementary statements. Such elementary statements are for example

- "Paris is the capital of France."
- "6 is a prime number."

An elementary statement is either true or false, but cannot be both. Each statement having the property of being either true or false is called a **proposition**. In the example, the first statement has the **truth value** TRUE, whereas the second statement has the truth value FALSE. The aim of propositional calculus is to characterize how these truth values can be inferred from elementary statements towards more complicated statements, for example towards "Paris is the capital of France, and 6 is a prime number". For two propositions A and B, the **conjunction** $A \wedge B$, the **disjunction** $A \vee B$, and the **complement** $\overline{A}$ are defined as follows:

$$A \wedge B \text{ is TRUE} \iff A \text{ is TRUE and } B \text{ is TRUE,} \tag{2.1}$$

$$A \vee B \text{ is TRUE} \iff A \text{ is TRUE or } B \text{ is TRUE,} \tag{2.2}$$

$$\overline{A} \text{ is TRUE} \iff A \text{ is FALSE.} \tag{2.3}$$

Predicate calculus. Predicate calculus is an extension of propositional calculus. The additional components are *predicates* and *quantifiers*. These concepts allow to describe circumstances that cannot be expressed by propositional calculus. Consider, for example, the property that two real numbers satisfy the inequality

$$x \cdot y \leq 5.$$

This inequality is *not* a proposition, as its truth value depends on the chosen values of the two variables x and y. If, however, any particular pair of numbers is substituted for x and y, then it becomes a proposition.

Predicates. Let $x_1, \ldots, x_n$ be variables with values in the domains S_1, $\ldots, S_n$. An expression $P(x_1, \ldots, x_n)$ involving variables $x_1, \ldots, x_n$ is a **predicate** if it becomes a proposition for each substitution of $x_1, \ldots, x_n$ by values in $S_1, \ldots, S_n$. Incidentally, a proposition can be interpreted as a predicate in zero variables.

Quantifiers. Predicate calculus allows one to express that a property holds *for all* objects, or that at least one object with a certain property *exists*. This is achieved by means of two *quantifiers*: the *universal quantifier* $\forall$ and the *existential quantifier* $\exists$. Thus, for a predicate $P(x_1, \ldots, x_n)$,

$$\forall x_i \ P(x_1, \ldots, x_n)$$

denotes a new predicate P' in the variables $x_1, \ldots, x_{i-1}, x_{i+1}, \ldots, x_n$. The predicate P' is true for a given substitution of values to its $n-1$ variables if this substitution makes the predicate P true for all possible substitutions of x_i by values in S_i. Analogously, the predicate

$$\exists x_i \ P(x_1, \ldots, x_n)$$

denotes a predicate P' in the variables $x_1, \ldots, x_{i-1}, x_{i+1}, \ldots, x_n$, which is true for a given substitution of its $n-1$ variables if this substitution makes P true for at least one possible substitution of x_i.

Example 2.1. Let $x_1, \ldots, x_n$ be real-valued variables. Let the predicate $P(x_1, x_2, x_3)$ be defined by

$$x_2 \geq 1 \ \wedge \ x_1 + x_2 - x_3 \geq 5, \tag{2.4}$$

and the predicate $P_1(x_1, x_3)$ be defined by

$$P_1(x_1, x_3) = \forall x_2 \ P(x_1, x_2, x_3).$$

Obviously, P_1 can be simplified to

$$x_1 - x_3 \geq 4.$$

The predicate

$$P_2(x_1, x_3) = \exists x_2 \ P_1(x_1, x_2, x_3)$$

is TRUE for all substitutions of x_1, x_3, as we can always find a suitable x_2 which satisfies condition (2.4).   ◇

2.2 Sets, Relations, and Functions

Sets. A **set** is a collection of objects. The objects in the collection are called **elements**. If the set is **finite**, i.e., it consists of only a finite number of elements, then it can be described by explicitly listing the elements, e.g.,

$$\{a, b, c\}.$$

The order in which the elements are enumerated does not matter. A second possible way to describe sets is by means of a **membership property**. To specify a set S in this way, we write

$$S = \{x : P(x)\},$$

where the predicate $P(x)$ is true if and only if the element x belongs to S. This method also works in case of infinite sets, e.g.,

$$S = \{x : x \text{ is a natural number divisible by 3 }\}.$$

Set membership of an element x in S is denoted by $x \in S$. The following sets are of particular importance, and therefore they are abbreviated by specific symbols:

$\mathbb{R}$: the set of real numbers,
$\mathbb{R}_0^+$: the set of non-negative real numbers,
$\mathbb{Z}$: the set of integers,
$\mathbb{N}$: the set of natural numbers.

If every element of a set S also belongs to a set T, then we say S is a **subset** of T and write $S \subset T$. Two sets S and T are equal if and only if $S \subset T$ and $T \subset S$. The **cardinality** of a set S is the number of its elements and is denoted by $|S|$ or $\#S$. The empty set which consists of zero elements is represented by the symbol $\emptyset$. A decomposition of a set S in non-empty, disjunct subsets is called a **partition** of S.

Operations on sets. The **Cartesian product** $S \times T$ of two sets S, T is defined by

$$S \times T = \{(x, y) : x \in S \text{ and } y \in T\}.$$

Thus, for example,

$$\{1, 3\} \times \{2, 4, 6\} = \{(1, 2), (1, 4), (1, 6), (3, 2), (3, 4), (3, 6)\}.$$

Here, the pairs (x, y) are **ordered pairs**, i.e., the order of the elements x, y within a pair (x, y) cannot be exchanged as in sets. The **union** $\cup$, **intersection** $\cap$, and **difference** $\backslash$ of sets are defined by

$$SUT = \{x : x \in S \text{ or } x \in T\},$$
$$S \cap T = \{x : x \in S \text{ and } x \in T\},$$
$$S \setminus T = \{x : x \in S \text{ but } x \notin T\}.$$

Power set. The **power set** of a set S, denoted by 2^S, is the set of subsets of S, i.e.,

$$2^S = \{R : R \subset S\}.$$

If, for example, $S = \{1, 2\}$, then

$$2^S = \{\emptyset, \{1\}, \{2\}, \{1, 2\}\}.$$

The notation 2^S originates from the fact that the cardinality of the power set of S is exactly $2^{|S|}$.

Relations. A (**binary**) **relation** R between two sets S and T is a subset of $S \times T$. With this subset R, the following predicate $x\,R\,y$ is associated:

$$x\,R\,y \iff (x, y) \in R.$$

If for given x, y this predicate is true, we say that x is in relation R with y. Very often, the two involved sets are identical; in this case we speak of a **relation on** S.

Equivalence relation. A relation R on a set S is called an **equivalence relation** if for all $x, y, z \in S$

- $x\,R\,x$ (**Reflexivity**),
- $x\,R\,y \implies y\,R\,x$ (**Symmetry**),
- $x\,R\,y \ \wedge \ y\,R\,z \implies x\,R\,z$ (**Transitivity**).

A relation R on a set S is called a **partial order** (**poset**) if it is reflexive, transitive, and **antisymmetric**. Here, antisymmetric means

$$x\,R\,y \ \wedge \ y\,R\,x \implies x = y.$$

A (**linear**) **order** is a partial order in which, for every pair $x \neq y$, either $x\,R\,y$ or $y\,R\,x$. Consequently, every set of elements can be arranged linearly in a way that is consistent with the linear order.

Example 2.2. 1. For each set S the subset relation $\subset$ defines a partial order on the power set 2^S.

2. The relation $\leq$ on the set of real numbers is a linear order.   ◇

Functions. A **function** f from a set S to a set T, written

$$f : S \to T,$$

assigns to every element $x \in S$ an element $f(x) \in T$. $f(x)$ is called the **image** of x under f. A function can be interpreted as a relation between the two sets in which each element of S appears as the first element in exactly one pair of the relation. The sets S and T are called the **domain** and **co-domain** of f, respectively. The set $f(S) = \{f(x) : x \in S\}$ is called the **range** or **image** of f.

A function $f : S \to T$ is called

- **surjective** or **onto**, if the range of f is equal to its co-domain, i.e., if $f(S) = T$,
- **injective** or **one-to-one**, if no two different elements of S have the same image under f, i.e., $f(x_1) = f(x_2)$ implies $x_1 = x_2$.
- **bijective** if f is both surjective and injective.

For a given function $f : S \to T$ and a subset $A \subset S$, the **image of A under** f is defined by $f(A) = \{f(x) : x \in A\}$. Conversely, for a subset $B \subset T$, the **inverse image of B under** f is $f^{-1}(B) = \{x \in S : f(x) \in B\}$.

2.3 Graphs

Directed graph. A (**directed**) **graph** G consists of a finite **vertex set** V and a set E of **edges** between the vertices. The set E is a subset of $V \times V$. The names *vertices* and *edges* are motivated by the pictorial representation we have in mind when speaking of a graph. For example, let $V = \{1, 2, 3, 4, 5\}$ and $E = \{(1, 2), (1, 3), (2, 3), (2, 4), (5, 2), (5, 3)\}$. Then G can be visualized by the diagram in Fig. 2.1 (a).

Here, two nodes u and v are connected by an edge starting in u if and only if $(u, v) \in E$. The edge set E defines a relation on the set of edges. Conversely, each binary relation R on a set S can be interpreted as a graph with vertex set S and edge set R.

Undirected graph. In case of an **undirected graph** the edges are considered as unordered pairs and therefore have no distinguished direction. The undirected graph which is associated with the directed graph in Fig. 2.1 (a) is depicted in Fig. 2.1 (b). The edge set of the undirected graph is $E = \{\{1, 2\},$

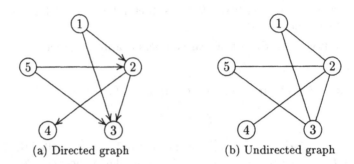

(a) Directed graph   (b) Undirected graph

Figure 2.1. Graphs

$\{1, 3\}, \{2, 3\}, \{2, 4\}, \{2, 5\}, \{3, 5\}\}$. Here, each edge is written in the form of a set $\{u, v\}$ to indicate the indifference in the order of u and v.

The **indegree** of a vertex v is the number of edges leading to v, i.e.,

$$indegree(v) = \#\{(u, v) \in E : u \in V\}.$$

Analogously, the **outdegree** of v is the number of edges starting in v,

$$outdegree(v) = \#\{(v, u) \in E : u \in V\}.$$

In case of undirected graphs, for each node the outdegree and the indegree coincide, and one uses the shorter term **degree**. A node is called a **sink** if it has outdegree 0. If the outdegree of v is bigger than 0, v is called an **internal node**. Analogously, a node is called a **root** if it has indegree 0.

If (u, v) is an edge, then u is called a **predecessor** of v, and v is called a **successor** of u. A **path** of length $k - 1$ is a sequence $v_1, \ldots, v_k$ of k nodes where v_{i+1} is a successor of v_i for all $1 \le i \le k-1$. If $v_1 = v_k$ the path is called **cyclic**. A graph is called an **acyclic graph** if there does not exist a cyclic path. A directed graph G is called **connected** if for every pair $(u, v) \in V$ there exist nodes $u = v_0, v_1, v_2, \ldots, v_k = v$ such that either (v_i, v_{i+1}) or (v_{i+1}, v_i) is an edge.

Tree. A graph is called **rooted** if there exists exactly one node with indegree 0, the **root**. A **tree** is a rooted acyclic graph in which every node but the root has indegree 1. This implies that in a tree, for every vertex v there exists a unique path from the root to v. The length of this path is called the **depth** of v. Figure 2.2 shows a tree with the depth values of its nodes. In the tree, a successor w of a node v is called **son** of v. v itself is called the **father** of w.

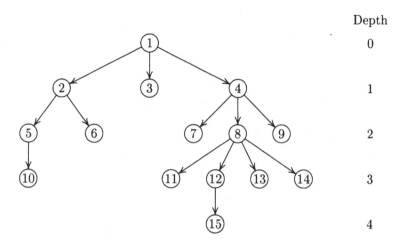

Figure 2.2. A tree with depth values

2.4 Algorithms and Data Structures

Half the battle is knowing what **problem** to solve. When initially approached, most problems have no simple, precise specification. In fact, even the word "problem" has very different meanings in different contexts. For our purposes, problems typically originate from questions with several parameters whose values have not yet been determined. A problem is defined by fixing the values of the parameters and setting conditions on admissible solutions.

Problem. A **decision problem** is a problem which has only two possible solutions: either Yes or No. An example of a decision problem is the task of deciding whether a given graph contains a cyclic path. An **optimization problem** is concerned with finding an *optimal* solution from a (possibly large) set of admissible solutions. Here, the quality of a solution is measured by means of a cost function. As it may sometimes be hard to find an optimal solution, one is often also interested in approximate solutions, i.e., admissible solutions whose quality closely come up to the quality of an optimal solution.

Algorithm. An **algorithm** is a description of a mechanical method for solving a given problem. It may have different forms, like a sequence of instructions or a computer program. In any case, an algorithm is expected to satisfy the following general properties:

Finiteness of description: The method can be described by means of a finite amount of text.

Effectiveness: Each single step of the description can be performed mechanically.

Termination: For each admissible input, the method stops after finitely many steps.

For a formal definition of the term *algorithm* very strict mathematical and theoretical models like **Turing machines** or λ-**calculus** are required. However, these and all other formalizations of the term *algorithm* have led to the same result. Based on this experience, the American logician A. Church stated the following thesis known as **Church's Thesis**:

> "Every function which is computable in the intuitive sense is
> Turing-computable."

In other words, this says that for each algorithmic method in arbitrary form, a Turing machine can be constructed that computes the same function.

Example 2.3. As an example of an algorithm and of the algorithm notation which we will use, we consider **depth first search** in a directed graph $G = (V, E)$. Starting from an initial node v_0, in systematic manner those nodes v are traversed for which there exists a path from v_0 to v. In depth first search, the order in which the nodes are traversed is defined as follows. After a node v is visited, all successor nodes of v are visited next, in a recursive manner. In order to guarantee that each node reachable from the initial node is only visited once, after the treatment of v the flag $mark[v]$ is set to 1. Pseudo code for depth first search is shown in Fig. 2.3. $\diamond$

Data structures. The question of how far algorithms can be performed efficiently depends strongly on the organization of the data being processed. A **data type** of a variable is the set of values that the variable can assume. For example, a variable of type *integer* contains values from the set $\mathbb{Z}$. An **abstract data type** is a mathematical model which considers the objects to be represented and the basis operations for manipulating these objects as a unity. To represent an abstract data type we use **data structures**, which are collections of variables, possibly of different data types, connected in various ways.

The quality of a data structure is judged from the viewpoint of how far it provides an optimal implementation of the abstract data type. In particular, a good data structure is expected to yield memory-efficient representation of the objects, and to enable efficient algorithms for solving the relevant basic operations. Data structures belong to the central research topics in computer science. In this book, we will see that even small modifications in complex data structures can change their algorithmic properties completely.

```
DFS_start(G = (V, E), v₀) {
/* Input: A graph G = (V, E), an initial node v₀ ∈ V */
/* Output: All nodes being reachable from v₀ in a depth first manner */
    For all nodes v ∈ V {
        mark[v] = 0;
    }
    DFS(G, v₀);
}

DFS(G = (V, E), v) {
    Output: "Node v";
    mark[v] = 1;
    For all successor nodes u of v {
        If (mark[u] = 0) {
            DFS(G, u);
        }
    }
}
```

Figure 2.3. Depth first search in a directed graph

2.5 Complexity of Algorithms

Algorithms will typically be judged according to their running time and space requirements. In complexity theory, these resource demands are measured in terms of the input size. This way, different algorithms for the same problems can be compared to each other.

Complexity. The **time complexity** $t_A(n)$ of an algorithm A denotes the maximal number of steps A needs for solving a problem instance with input size n. Analogously, the **space complexity** $s_A(n)$ denotes the maximal number of memory cells used to solve a problem with input size n.

Asymptotic complexity analysis. Often, it is impossible to determine the exact complexity of an algorithm A. However, then one is at least interested in the rate of growth of the functions $t_A(n)$ and $s_A(n)$. Good estimations of this growth serve as criteria to judge the quality of the algorithm. When describing the growth, it is useful, for example, to neglect constant factors. This approach is justified as the constant factors depend quite strongly on such technical implementation details as the chosen programming language. In the analysis, the influence of those technical details should not prevent one from recognizing the essentials. Moreover, it seems reasonable not to neglect only constant factors, but to concentrate in the complexity analysis solely on the dominant terms of the complexity functions. Here, one speaks of **asymptotic analysis**, a widespread analysis technique in computer science.

Indeed, without this technique many basic insights in computer science would not have been gained.

For asymptotically characterizing upper bounds of a complexity function $f : \mathbb{N} \to \mathbb{R}_0^+$, one uses the notation

$$f(n) = \mathcal{O}(g(n))$$

if there exist two constants $c, n_0 \in \mathbb{N}$ such that for all $n \geq n_0$

$$f(n) \leq c \cdot g(n).$$

In fact, besides constant factors, this notation also neglects terms which are of secondary importance for the growth of the investigated function. We read "f is of order at most g".

Example 2.4. $f(n) = \mathcal{O}(1)$ means $f(n) \leq c$ for a constant c.
$f(n) = n^{\mathcal{O}(1)}$ expresses that f is bounded by a polynomial in n.   ◇

If we are interested in lower bounds for a complexity function $f(n)$, then the following notation is used. We say

$$f(n) = \Omega(g(n)),$$

read "f is of order at least g", if there exist two constants $c, n_0 \in \mathbb{N}$, such that for all $n \geq n_0$

$$f(n) \geq c \cdot g(n).$$

Furthermore, we write

$$f(n) = \Theta(g(n))$$

if $f(n) = \mathcal{O}(g(n))$ and $g(n) = \mathcal{O}(f(n))$, i.e., if the rates of growth are identical for f and g.

In many applications, only those algorithms are practical whose running time is bounded from above by a polynomial. An important – maybe *the* most important – task in theoretical computer science is to answer the question of when and for which problems such algorithms exist. In the following, we will concentrate our description on decision problems.

The complexity class P. The class **P** (polynomial time) denotes the set of all decision problems that can be solved by means of polynomial time algorithms. Obviously, the space complexity of a polynomial time algorithm is also polynomially bounded.

The complexity class NP. The subsequently formally defined class **NP** (nondeterministic polynomial) denotes problems that can at least be solved efficiently in a nondeterministic way. In contrast to the deterministic case, where in each situation exactly one action can be performed, in the nondeterministic case, a variety of different actions are possible.

If, for example, one considers the search for a proof of a mathematical theorem, then in case of a wrong claim there does not exist a proof at all. If, however, the claim is true, then in general different proofs can be provided. For proving the correctness of a theorem it is merely important that *at least one* proof can be given. Of course, finding a proof can be arbitrarily difficult. But if a proof is presented, then it is no longer difficult to understand it and to accept the claim. In complexity theory, such a proof is called a **witness** or **certificate**.

Definition 2.5. *A decision problem A belongs to* **NP** *if there exist a polynomial p and a polynomial time algorithm A which computes, for each input x and each possible certificate y of length at most $p(|x|)$, a value $t(x, y)$ such that:*

1. *If the answer to the input x is "No", then $t(x, y) = 0$ for all possible certificates.*
2. *If the answer to the input x is "Yes", then $t(x, y) = 1$ for at least one certificate.*

"P $\overset{?}{=}$ NP". Obviously, the class **P** is contained in the class **NP**. The most important open problem in modern complexity theory is the question

$$\text{"P} \overset{?}{=} \text{NP"}.$$

The importance of this question originates from the fact that there are numerous practically relevant tasks for which polynomial algorithms are not known, but for which membership in the class **NP** can be proven. Without a clarification of the question "**P** $\overset{?}{=}$ **NP**" it is not possible of deciding whether these tasks cannot be solved in polynomial time at all, or whether such algorithms have only not been found so far.

Nearly all experts in the area of complexity theory conjecture that the classes **P** and **NP** are different. This conjecture is supported by a series of results in the investigation of the subsequently defined concepts of *polynomial time reduction* and **NP**-*complete problems*. **NP**-complete problems represent the "hardest" problems within the class **NP**. It is proven that if *any* **NP**-complete problem can be solved in polynomial time, then so can all other problems in **NP**, and it follows that **P** = **NP**.

Definition 2.6. *Let A and B two problems. A is called **polynomial time reducible** to B, if, under the assumption that arbitrary instances of B can be solved in constant time, a polynomial time algorithm for A exists. We write $A \leq_P B$.*

Now the following lemma follows easily:

Lemma 2.7. *If $A \leq_P B$ and $B \in \mathbf{P}$ then $A \in \mathbf{P}$.*   □

Definition 2.8. *1. A problem A is called **NP-hard** if all $B \in \mathbf{NP}$ satisfy: $B \leq_P A$.*

  *2. A problem A is called **NP-complete**, if both A is **NP**-hard and $A \in \mathbf{NP}$.*

Theorem 2.9. *Let A be an **NP**-complete problem. Then the following holds:*

  1. If $A \in \mathbf{P}$ then $\mathbf{P} = \mathbf{NP}$.

  *2. If $A \notin \mathbf{P}$ then all **NP**-complete problems B satisfy $B \notin \mathbf{P}$.*

Proof. Let $A \in \mathbf{P}$, and let B be an arbitrary problem in $\mathbf{NP}$. As in particular A is **NP**-hard, we have $B \leq_P A$. Then Lemma 2.7 implies $B \in \mathbf{P}$. As B was chosen arbitrarily from $\mathbf{NP}$, we have $\mathbf{P} = \mathbf{NP}$.

Now let $A \notin \mathbf{P}$, and let B be an **NP**-complete problem with $B \in \mathbf{P}$. According to the first statement of the theorem we have $\mathbf{P} = \mathbf{NP}$, and hence $A \in \mathbf{P}$ in contradiction to the assumption.   □

Example 2.10. (1) A man has invested all his money in 50 gold nuggets. The weight of the i-th nugget is g_i ounces. For simplicity, we assume that $g_1, g_2, \ldots, g_{50}$ are natural numbers. When the man dies, his last will contains the following condition: If his riches can be split into two parts of equal weight, then each of his daughters should obtain one of these parts. If instead there is no such partition, then all his gold should go to the church.

The general question therefore is: Is it possible to divide a set of n natural numbers $\{g_1, \ldots, g_n\}$ in two subsets such that the sums over each subset are equal ? This problem is called **PARTITION** and is **NP**-complete. The straightforward trial-and-error method is only practical for small n, as it requires testing of all 2^n possibilities in the worst case. Note that if someone gives you a solution, you can verify this solution very easily.

(2) The following problem, called **HAMILTONIAN CIRCUIT**, is also **NP**-complete. Given an undirected graph with n nodes, does there exist a cyclic path of length n through the graph which traverses each node in the graph exactly once ?   ◇

Definition 2.11. *The* **complementary** *problem* $\overline{A}$ *of a decision problem* A *results from* A *by negating all answers.*

A decision problem A *is called* **co-NP-complete**, *if the complementary problem* $\overline{A}$ *is* **NP**-*complete.*

Example 2.12. The following problem is **co-NP**-complete:

Input: An undirected graph with n nodes.

Output: "Yes", if there *does not* exist a cyclic path of length n which traverses each node in the graph exactly once. "No", otherwise.   ◇

2.6 Hashing

Hashing. The term **hashing** denotes data storage and retrieval mechanisms which compute the addresses of the data records from corresponding keys. Formally, a hash function

$$h : X \rightarrow A$$

has to be defined which maps the set of keys X to a set of addresses A. A data record with a key $x \in X$ is then stored at the address $h(x)$. Each subsequent retrieval request for this data record consists solely of computing the hash value $h(x)$ and searching at this address.

In the design of hash techniques two important aspects have to be considered:

Collisions: It has to be determined how the case of two keys x and y with $x \neq y$ and identical hash value $h(x) = h(y)$ is handled.

Choice of hash function: The hash function h should be easy to compute and should distribute the set of keys X as even and as randomly as possible (in order to keep the number of collisions small).

Collision list. An easy way to treat collisions is to link data objects with identical hash values together in a linear list, the so-called **collision list**. Searching for a data record with key x then requires computing $h(x)$, and subsequently searching sequentially for x in the collision list corresponding to $h(x)$.

Example 2.13. Let X be the set of all strings, and let $A = \{0, \dots, 99\}$. For a string $x = x_1 \dots x_k$, we define the corresponding hash value by

$$h(x) = h(x_1 \dots x_k) = \sum_{i=1}^{k} \mathrm{asc}(x_i) \pmod{100},$$

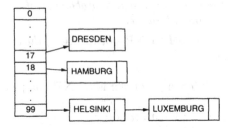

Figure 2.4. Hash table with collision lists

where asc(x_i) denotes the ASCII value of symbol x_i (e.g., asc('A') = 65, ... , asc('Z') = 90).

Then h(HELSINKI) = $72 + 69 + 76 + 83 + 73 + 78 + 75 + 73$ (mod 100) = 99, h(DRESDEN) = 17, h(LUXEMBURG) = 99, h(HAMBURG) = 18. If collision lists are used, the resulting hash table is depicted in Fig. 2.4.   ◇

2.7 Finite Automata and Finite State Machines

Finite automata. Finite automata provide a mathematical model for describing computation processes whose memory requirements are bounded by a constant. A finite automaton consists of a finite set of (memory) states and a specification of the state transitions. The state transitions themselves depend on the current state and on the current input.

State transition. As an easy example, let us consider a counter with reset state q_0. Whenever the input is 1, the counter periodically traverses the states q_0, q_1, q_2, q_3. If the input is 0, then the counter immediately returns to the reset state q_0. The behavior of the counter can be precisely described by a **transition table** or by a **state diagram**, see Fig. 2.5. A state diagram is a graph whose nodes are labeled by the names of the states and whose edges characterize the state transitions. The labels of the edges identify the inputs on which the transitions takes place.

Formally, a finite state machine is defined by the following components:

- the **state set** Q,
- the **input alphabet** I,
- the **next-state function** $\delta : Q \times I \to Q$,
- and the **initial state** q_0.

Depending on the context, a subset of the states may be distinguished as the set of final states.

State	Input	Successor state
q_0	0	q_0
q_0	1	q_1
q_1	0	q_0
q_1	1	q_2
q_2	0	q_0
q_2	1	q_3
q_3	0	q_0
q_3	1	q_0

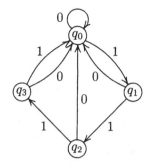

Figure 2.5. Transition table and state diagram of a finite automaton

Finite state machine. A sequential system or a **finite state machine** is a finite automaton which is extended by an output behavior. Each transition in the finite state machine is associated with an output symbol. Whenever a transition takes place, the associated output symbol is sent to an output device. In Fig. 2.6, a finite state machine is shown which has the same internal behavior as the counter in Fig. 2.5. Additionally, the output is 1 whenever the machine enters the reset state q_0, and the output is 0 for all other transitions.

Formally, a finite state machine is a finite automaton which is extended by the following two components:

- the output alphabet O, and
- the output function $\lambda : Q \times I \to O$.

State	Input	Successor state	Output
q_0	0	q_0	1
q_0	1	q_1	0
q_1	0	q_0	1
q_1	1	q_2	0
q_2	0	q_0	1
q_2	1	q_3	0
q_3	0	q_0	1
q_3	1	q_0	1

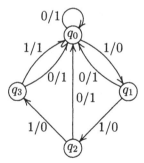

Figure 2.6. Transition table and state diagram of a finite state machine

2.8 References

The topics of this chapter all belong to the fundamentals of computer science and discrete mathematics, so we cite merely a small collection of relevant textbooks. Detailed information on the topics of discrete mathematics can be found in [Big90]. The design and analysis of algorithms and data structures is covered in [AHU74] and [CLR90]. Introductions to the theory of **NP**-completeness can be found in [GJ78, Pap94]. In particular, the compendium [GJ78] additionally contains a complexity theoretical classification of many problems, in particular with regard to **NP**-completeness.

A standard work concerning finite automata is [HU78]. Finally, the books [Mic94] and [HS96] excellently cover algorithmic aspects and the significance of finite state machines in VLSI design.

Part I

Data Structures for Switching Functions

3. Boolean Functions

*Pure mathematics was discovered by Boole in a work
which he called "The Laws of Thought".*

Bertrand Russell (1872–1970)

The basic components of computers and other digital systems are **circuits**. Circuits consist of wires connecting one or several voltage input ports with one or several output ports. The simplified model for describing the relation between input and output voltage merely starts from two well-distinguished signals: either there is a voltage, indicated by the value 1, or there is none, indicated by the value 0. Of course, in a technical realization, there will be deviations from these nominal values. However, as long as these deviations do not exceed certain limits, they do not affect the basic functionality of the circuit. Hence, in fact, the statement that a signal is two-valued or *binary* simply means that the value of this signal is within one of two non-overlapping continuous ranges.

VLSI circuits consist of a complex combination of a limited number of basic elements, so-called **gates**, which perform simple logical operations. These operations do not depend on the *exact* value of the input signal, but only on the corresponding range. For modeling the binary signals one uses binary variables, i.e., variables with values in the domain $\{0, 1\}$. If the binary variables $x_1, \ldots, x_n$ denote input signals of a circuit, then the output signal y (which is determined uniquely by the input variables) can be described in terms of *switching functions* $y = f(x_1, \ldots, x_n)$. Obviously, those switching functions are defined on the set of bit vectors $\{0, 1\}^n$. In 1936, this "calculus of switching circuits" was developed by C. E. Shannon. He showed how the basic rules of classical logic and elementary set theory, formulated by G. Boole in his *Laws of Thought* in 1854, could be used to describe and analyze circuits, too.

The purpose of Section 3.1 and Section 3.2 of this chapter is to review fundamental definitions and basic properties of Boolean algebra and Boolean functions. In Section 3.3, we consider the special case of switching functions. At first glance, these functions, which map tuples of variables to one of only two possible values, may look very simple. However, the almost unlimited possibilities for combining switching functions through many stages of modern engineering systems give Boolean analysis its own typical complexity in theory and practice.

3.1 Boolean Algebra

Definition 3.1. *A set A together with two distinguished, distinct elements* **0** *and* **1***, two binary operations* $+$*,* $\cdot$*, and a unary operation* $^-$ *is called a* **Boolean algebra** $(A; +, \cdot, ^-, 0, 1)$ *if for all* $a, b, c \in A$ *the following axioms are satisfied:*

Commutative law:

$$a + b = b + a \quad and \quad a \cdot b = b \cdot a.$$

Distributive law:

$$a \cdot (b + c) = (a \cdot b) + (a \cdot c) \quad and \quad a + (b \cdot c) = (a + b) \cdot (a + c).$$

Identity law:

$$a + 0 = a \quad and \quad a \cdot 1 = a.$$

Complement law:

$$a + \bar{a} = 1 \quad and \quad a \cdot \bar{a} = 0.$$

The set A is called the **carrier***. The distinguished elements* **0** *and* **1** *are called the* **zero element** *and the* **one element***, respectively.*

In the following, we shall only be concerned with **finite** Boolean algebras, i.e., Boolean algebras whose carrier is finite. Following the usual parentheses conventions, the unary operation $^-$ couples more tightly than $\cdot$, and $\cdot$ couples more tightly than $+$. As usual in arithmetics, we may omit the symbol $\cdot$, i.e., $a \cdot b$ is also written ab.

Example 3.2. (1) If 2^S denotes the power set of a set S, and if for each set $A \subset S$ the term $\bar{A}$ denotes the set $S \setminus A$, then

$$(2^S; \cup, \cap, ^-, \emptyset, S)$$

is a Boolean algebra, the so-called **set algebra** of S.

(2) For a natural number $n > 1$ let T_n be the set of divisors of n. Then

$$(T_n; \operatorname{lcm}(\cdot, \cdot), \gcd(\cdot, \cdot), (\cdot)^{-1}, 1, n)$$

constitutes a Boolean algebra, where $\mathrm{lcm}(\cdot, \cdot)$ and $\gcd(\cdot, \cdot)$ denote the least common multiple and the greatest common divisor, and $(\cdot)^{-1}$ is the reciprocal value in the sense $(x)^{-1} = n/x$.

(3) Let $\mathcal{B} = (\mathcal{A}; +, \cdot, ^-, \mathbf{0}, \mathbf{1})$ be a Boolean algebra. The set $F_n(\mathcal{A})$ of all functions from $\mathcal{A}^n$ to $\mathcal{A}$ constitutes a Boolean algebra $\mathcal{F}(\mathcal{A}) = (F_n(\mathcal{A}); +, \cdot, ^-, \mathbf{0}, \mathbf{1})$ with respect to the following operations and elements:

$$
\begin{aligned}
f + g : \mathcal{A}^n &\longrightarrow \mathcal{A} & a = (a_1, \ldots, a_n) &\mapsto f(a) + g(a), \\
f \cdot g : \mathcal{A}^n &\longrightarrow \mathcal{A} & a = (a_1, \ldots, a_n) &\mapsto f(a) \cdot g(a), \\
\overline{f} : \mathcal{A}^n &\longrightarrow \mathcal{A} & a = (a_1, \ldots, a_n) &\mapsto \overline{f(a)}, \\
\mathbf{0} : \mathcal{A}^n &\longrightarrow \mathcal{A} & (a_1, \ldots, a_n) &\mapsto \mathbf{0}, \\
\mathbf{1} : \mathcal{A}^n &\longrightarrow \mathcal{A} & (a_1, \ldots, a_n) &\mapsto \mathbf{1}.
\end{aligned}
$$

$\diamond$

Investigating Boolean algebras, one realizes that identities and properties come in pairs. Briefly stated, the so-called **principle of duality** expresses that for each valid equation over a Boolean algebra, the dual equation is also valid.

Definition 3.3. *Let G denote an equation over a Boolean algebra $\mathcal{B} = (\mathcal{A}; +, \cdot, ^-, \mathbf{0}, \mathbf{1})$. The equation G' resulting from G by systematically exchanging the two operations $+$ and $\cdot$, and by exchanging the two distinguished elements $\mathbf{0}$ and $\mathbf{1}$, is called the* **dual equation** *of G.*

For example, the dual equation of the axiom $a + \mathbf{0} = a$ is the equation $a \cdot \mathbf{1} = a$.

Theorem 3.4. (Principle of duality)
Let G' be the dual equation of G. If G is a true proposition in the theory of Boolean algebra, then this also holds for G'.

Proof. Considering the underlying axioms of a Boolean algebra, one immediately recognizes that the set of axioms does not change, when $\mathbf{0}$ and $\mathbf{1}$, and $+$ and $\cdot$, respectively, are exchanged.

Consequently, for each equation which can be deduced from the axioms of the Boolean algebra, the dual proposition can also be deduced. In each deduction step, one merely uses the dual of the justifying axiom. $\qquad\square$

By explicitly exploiting the principle of duality, propositions in Boolean algebra are always formulated in pairs. A proof, however, is only required for one of the two versions.

Theorem 3.5. (Computation rules)
For any elements $a, b, c \in \mathcal{A}$ in a Boolean algebra $(\mathcal{A}; +, \cdot, ^-, \mathbf{0}, \mathbf{1})$ we have the computation rules shown in Fig. 3.1.

Idempotence:

$$a + a = a \text{ and } a \cdot a = a,$$

Properties of 1 and 0:

$$a + 1 = 1 \text{ and } a \cdot 0 = 0,$$

Absorption:

$$a + (a \cdot b) = a \text{ and } a \cdot (a + b) = a,$$

Associativity:

$$a + (b + c) = (a + b) + c \text{ and } a \cdot (b \cdot c) = (a \cdot b) \cdot c,$$

DeMorgan's rules:

$$\overline{a + b} = \overline{a} \cdot \overline{b} \text{ and } \overline{a \cdot b} = \overline{a} + \overline{b},$$

Involution:

$$\overline{\overline{a}} = a.$$

Figure 3.1. Computation rules in a Boolean algebra

Proof. Idempotence:

$$
\begin{aligned}
a &= a + 0 && \text{(identity)} \\
&= a + (a \cdot \overline{a}) && \text{(complementation)} \\
&= (a + a) \cdot (a + \overline{a}) && \text{(distributivity)} \\
&= (a + a) \cdot 1 && \text{(complementation)} \\
&= a + a && \text{(identity)}
\end{aligned}
$$

Particular property of 1:

$$
\begin{aligned}
a + 1 &= (a + 1) \cdot 1 && \text{(identity)} \\
&= (a + 1) \cdot (a + \overline{a}) && \text{(complementation)} \\
&= a + (1 \cdot \overline{a}) && \text{(distributivity)} \\
&= a + \overline{a} && \text{(commutativity, identity)} \\
&= 1 && \text{(complementation)}
\end{aligned}
$$

Absorption:

$$
\begin{aligned}
a + (a \cdot b) &= (a \cdot 1) + (a \cdot b) && \text{(identity)} \\
&= a \cdot (1 + b) && \text{(distributivity)} \\
&= a \cdot 1 && \text{(commutativity, property of 1)} \\
&= a && \text{(identity)}
\end{aligned}
$$

Associativity:

Let $s_l = a + (b + c)$ denote the left side and $s_r = (a + b) + c$ denote the right side of the equation to be proved. First we show $a \cdot s_l = a \cdot s_r$:

$$\begin{aligned}
a \cdot s_l &= a \cdot (a + (b + c)) \\
&= (a \cdot a) + (a \cdot (b + c)) \quad \text{(distributivity)} \\
&= a + (a \cdot (b + c)) \qquad \text{(idempotence)} \\
&= a \qquad\qquad\qquad\quad \text{(absorption)}
\end{aligned}$$

Analogously, $a \cdot s_r = a$ and, hence, $a \cdot s_l = a \cdot s_r$. Now we prove $\bar{a} \cdot s_l = \bar{a} \cdot s_r$:

$$\begin{aligned}
\bar{a} \cdot s_l &= \bar{a} \cdot (a + (b + c)) \\
&= (\bar{a} \cdot a) + (\bar{a} \cdot (b + c)) \quad \text{(distributivity)} \\
&= \mathbf{0} + (\bar{a} \cdot (b + c)) \qquad \text{(complementation)} \\
&= \bar{a} \cdot (b + c) \qquad\qquad \text{(commutativity, identity)}
\end{aligned}$$

Analogously, $\bar{a} \cdot s_r = \bar{a} \cdot (b + c)$ and, hence, $\bar{a} \cdot s_l = \bar{a} \cdot s_r$.
Altogether we obtain:

$$\begin{aligned}
s_l &= s_l \cdot \mathbf{1} \qquad\qquad\qquad \text{(identity)} \\
&= s_l \cdot (a + \bar{a}) \qquad\qquad \text{(complementation)} \\
&= (s_l \cdot a) + (s_l \cdot \bar{a}) \quad \text{(distributivity)} \\
&= (s_r \cdot a) + (s_r \cdot \bar{a}) \quad \text{(commutativity, } a \cdot s_l = a \cdot s_r) \\
&= s_r \cdot (a + \bar{a}) \qquad\qquad \text{(distributivity)} \\
&= s_r \cdot \mathbf{1} \qquad\qquad\qquad \text{(complementation)} \\
&= s_r \qquad\qquad\qquad\quad \text{(identity)}
\end{aligned}$$

DeMorgan's rules:

First we show that the complement $\bar{a}$ of an element is already uniquely determined by satisfying the two complement laws. For this reason, assume that two elements b and c satisfy the complement laws with respect to a, i.e.,

$$a + b = \mathbf{1}, \quad a \cdot b = \mathbf{0}, \quad a + c = \mathbf{1}, \quad a \cdot c = \mathbf{0}.$$

This implies

$$\begin{aligned}
b &= b \cdot (a + c) + a \cdot c = b \cdot a + b \cdot c + a \cdot c \\
&= b \cdot a + (b + a) \cdot c = \mathbf{0} + \mathbf{1} \cdot c = c.
\end{aligned}$$

Now we show that $a + b$ and $\bar{a} \cdot \bar{b}$ satisfy the two complement laws:

$$\begin{aligned}
(a + b) + (\bar{a} \cdot \bar{b}) &= ((a + b) + \bar{a}) \cdot ((a + b) + \bar{b}) = ((a + \bar{a}) + b) \cdot (a + (b + \bar{b})) \\
&= (\mathbf{1} + b) \cdot (a + \mathbf{1}) = \mathbf{1} \cdot \mathbf{1} = \mathbf{1}. \\
(a + b) \cdot (\bar{a} \cdot \bar{b}) &= a \cdot \bar{a} \cdot \bar{b} + b \cdot \bar{a} \cdot \bar{b} = \mathbf{0} + \mathbf{0} = \mathbf{0}.
\end{aligned}$$

Involution: Due to

$$\bar{a} + a = a + \bar{a} = 1,$$
$$\bar{a} \cdot a = a \cdot \bar{a} = 0,$$

as in the proof of DeMorgan's rules, we can deduce that a is the complement of $\bar{a}$, i.e., $\bar{\bar{a}} = a$.   □

The last theorem of this section, which will be stated without proof, expresses that for each finite Boolean algebra there exists a set algebra of exactly the same structure.

Definition 3.6. *Two Boolean algebras* $\mathcal{B} = (\mathcal{A}; +, \cdot, ^-, 0, 1)$ *and* $\mathcal{B}' = (\mathcal{A}'; +', \cdot', ^{-'}, 0', 1')$ *are called* **isomorphic** *if there exists a bijective mapping* $\phi : \mathcal{A} \to \mathcal{A}'$ *such that for all* $a, b \in \mathcal{A}$:

$$\phi(a + b) = \phi(a) +' \phi(b),$$
$$\phi(a \cdot b) = \phi(a) \cdot' \phi(b),$$
$$\phi(\bar{a}) = \overline{\phi(a)}',$$
$$\phi(0) = 0',$$
$$\phi(1) = 1'.$$

Theorem 3.7. (Stone's representation theorem)
Every finite Boolean algebra is isomorphic to the set algebra of some finite set.   □

3.2 Boolean Formulas and Boolean Functions

Boolean functions are particular functions which can be described in terms of expressions over a Boolean algebra, so-called *Boolean formulas*.

Definition 3.8. *Let* $\mathcal{B} = (\mathcal{A}; +, \cdot, ^-, 0, 1)$ *be a Boolean algebra. An expression consisting of* n *variable symbols* $x_1, \ldots, x_n$, *the symbols* $+, \cdot, ^-$, *and the elements of* $\mathcal{A}$ *is called an* n-**variable Boolean formula** *if it satisfies the following recursive rules:*

1. *The elements of* $\mathcal{A}$ *are Boolean formulas.*
2. *The variable symbols* $x_1, \ldots, x_n$ *are Boolean formulas.*
3. *If* F *and* G *are Boolean formulas, then so are*
   a) $(F) + (G)$,
   b) $(F) \cdot (G)$, *and*
   c) $\overline{(F)}$.

4. An expression is a Boolean formula if and only if it can be produced by means of finitely many applications of the rules 1, 2, and 3.

For the subsequent treatment, we relax this definition slightly. We still call an expression a Boolean formula if it can be derived from a Boolean formula by removing a pair of parentheses, and if this process (by means of the usual parentheses convention) still allows us to recognize the original meaning uniquely. Hence, the expressions $(x_1) + (a)$ and $x_1 + a$ are equivalent in this context.

In the above definition, Boolean formulas are merely regarded as formal strings of symbols. In order to transform these formulas into functions, the symbols $+, \cdot, ^-$ are interpreted as Boolean operations, and the variable symbols x_i are considered as input variables that can be substituted by elements of the Boolean algebra. To be more specific, let us investigate a Boolean algebra $\mathcal{B} = (\mathcal{A}; +, \cdot, ^-, 0, 1)$ and a formula F over this algebra. F **induces** the function f_F,

$$f_F : \mathcal{A}^n \longrightarrow \mathcal{A}, \quad a = (a_1, \ldots, a_n) \mapsto f_F(a_1, \ldots, a_n).$$

Here, $f_F(a_1, \ldots, a_n)$ denotes the element of $\mathcal{A}$ which results from substituting the variable symbols x_i in F by the elements $a_i \in \mathcal{A}$, and, according to the sequence being defined by F, performing the Boolean operations on these elements.

Definition 3.9. *Let $\mathcal{B} = (\mathcal{A}; +, \cdot, ^-, 0, 1)$ be a Boolean algebra. An n-variable function $f : \mathcal{A}^n \to \mathcal{A}$ is called a* **Boolean function** *if it is induced by a Boolean formula F. We say that formula F* **represents** *the function f. The set of all Boolean functions over $\mathcal{B}$ is denoted by $P_n(\mathcal{B})$.*

Example 3.10. Let us consider the set algebra $\mathcal{B} = \{\{1, 2\}, \cup, \cap, ^-, \emptyset, \{1, 2\}\}$ of the set $\{1, 2\}$ and the formula

$$F = x_1 + \overline{x_1} \cdot x_2.$$

The tabular representation of the induced Boolean function f_F is shown in Fig. 3.2. $\diamond$

Let $\mathcal{B} = (\mathcal{A}; +, \cdot, ^-, 0, 1)$ be a Boolean algebra. The set $P_n(\mathcal{B})$ of all n-variable Boolean functions over the algebra $\mathcal{B}$ is a subset of the set $F_n(\mathcal{A})$ of all functions from $\mathcal{A}^n$ to $\mathcal{A}$. According to Example 3.2, the set $F_n(\mathcal{A})$ with respect to the operations $+, \cdot, ^-$ (adjusted canonically to functions) is a Boolean algebra. Of course, in the same way the operations $+, \cdot, ^-$ of $\mathcal{B}$ induce operations on $P_n(\mathcal{B})$. In the following, we state the important property that the subset $P_n(\mathcal{B})$ is closed under these operations, and that it constitutes a Boolean algebra, too, a subalgebra of the Boolean algebra of all n-variable functions.

x_1	x_2	f_F
$\emptyset$	$\emptyset$	$\emptyset$
$\emptyset$	$\{1\}$	$\{1\}$
$\emptyset$	$\{2\}$	$\{2\}$
$\emptyset$	$\{1,2\}$	$\{1,2\}$
$\{1\}$	$\emptyset$	$\{1\}$
$\{1\}$	$\{1\}$	$\{1\}$
$\{1\}$	$\{2\}$	$\{1,2\}$
$\{1\}$	$\{1,2\}$	$\{1,2\}$

x_1	x_2	f_F
$\{2\}$	$\emptyset$	$\{2\}$
$\{2\}$	$\{1\}$	$\{1,2\}$
$\{2\}$	$\{2\}$	$\{2\}$
$\{2\}$	$\{1,2\}$	$\{1,2\}$
$\{1,2\}$	$\emptyset$	$\{1,2\}$
$\{1,2\}$	$\{1\}$	$\{1,2\}$
$\{1,2\}$	$\{2\}$	$\{1,2\}$
$\{1,2\}$	$\{1,2\}$	$\{1,2\}$

Figure 3.2. Tabular representation of a Boolean function

Definition 3.11. *Let* $\mathcal{B} = (\mathcal{A}; +, \cdot, ^-, 0, 1)$ *and* $\mathcal{B}' = (\mathcal{A}'; +, \cdot, ^-, 0, 1)$ *be two Boolean algebras with the same operations and the same zero and one elements.* $\mathcal{B}$ *is called a* **subalgebra** *of* $\mathcal{B}'$ *if the carrier* $\mathcal{A}$ *is a subset of the carrier* $\mathcal{A}'$.

Theorem 3.12. *Let* $\mathcal{B} = (\mathcal{A}; +, \cdot, ^-, 0, 1)$ *be a Boolean algebra. The set* $P_n(\mathcal{B})$ *of all n-variable Boolean functions over* $\mathcal{B}$ *with respect to the operations and constants*

$$
\begin{aligned}
+ \quad &: \quad \text{addition of functions,} \\
* \quad &: \quad \text{multiplication of functions,} \\
^- \quad &: \quad \text{complement operation on functions,} \\
\underline{0} \quad &: \quad \text{zero function,} \\
\underline{1} \quad &: \quad \text{one function,}
\end{aligned}
$$

defines a Boolean algebra which is a subalgebra of the Boolean algebra $(F_n(\mathcal{A}); +, \cdot, ^-, \underline{0}, \underline{1})$ *of all n-variable functions over* $\mathcal{B}$. $\square$

It is an interesting fact that the relationship between Boolean formulas and Boolean functions is not one-to-one: many different formulas represent the same Boolean functions. An important and central task in many applications of Boolean algebra is to find "good" formulas – according to problem-specific quality criteria – for representing certain concrete Boolean functions under investigation.

Example 3.13. The Boolean formulas $(x_1 + x_2) \cdot (x_3 + x_2 \cdot \overline{x_4} + \overline{x_1} \cdot x_2)$ and $x_3 \cdot (x_1 + x_2) + x_2 \cdot (\overline{x_1} + \overline{x_4})$ represent the same Boolean function over all Boolean algebras, as

$$
\begin{aligned}
&(x_1 + x_2) \cdot (x_3 + x_2 \cdot \overline{x_4} + \overline{x_1} \cdot x_2) \\
&= x_1 \cdot x_3 + x_1 \cdot x_2 \cdot \overline{x_4} + x_2 \cdot x_3 + x_2 \cdot \overline{x_4} + \overline{x_1} \cdot x_2 \\
&= x_1 \cdot x_3 + x_2 \cdot x_3 + x_2 \cdot \overline{x_4} + \overline{x_1} \cdot x_2 \\
&= x_3 \cdot (x_1 + x_2) + x_2 \cdot (\overline{x_1} + \overline{x_4}).
\end{aligned}
$$

If the quality criterion is to minimize the number of variable symbols occurring in the formula, the last formula will be preferred to the first one.

$\diamond$

Two n-variable Boolean formulas over a Boolean algebra $\mathcal{B}$ are said to be **equivalent** if for all 2^n input elements in $\{0,1\}^n$ the function values of both functions coincide.

Theorem 3.14. *Let* $\mathcal{B} = (\mathcal{A}, +, \cdot, ^-, 0, 1)$ *be a Boolean algebra. If two Boolean functions* f *and* g *are equivalent, then they are identical, i.e., their function values coincide for all inputs.*

Proof. Let F be a Boolean formula representing f. Using the stated axioms and identities as well as the definition of Boolean expressions, it can be shown that the formula can be expanded to the following sum of products:

$$F = \sum_{(e_1,\dots,e_n) \in \{0,1\}^n} a(e_1,\dots,e_n) \cdot x_1^{e_1} \cdot \dots \cdot x_n^{e_n},$$

where $x_i^1 := x_i$, $x_i^0 := \overline{x_i}$, and $a(e_1,\dots,e_n) \in \mathcal{A}$. Hence, the function f is defined uniquely by the coefficients $a(e_1,\dots,e_n)$. The coefficients $a(e_1,\dots,e_n)$ themselves only depend on the function values of f for the inputs $(e_1,\dots,e_n) \in \{0,1\}^n$. In the same way, the 2^n coefficients uniquely describing the function g only depend on the function values for the inputs $(e_1,\dots,e_n) \in \{0,1\}^n$. Hence, if f and g are equivalent, then they are identical. $\square$

As a consequence of the last proposition and its proof, the number of different n-variable Boolean functions can be determined. As the function values of all 2^n inputs $\{0,1\}^n$ can be chosen freely, there are $\#\mathcal{A}^{2^n}$ n-variable Boolean functions, where $\#\mathcal{A}$ is the cardinality of the carrier. However, the number of *all* functions $\mathcal{A}^n \longrightarrow \mathcal{A}$ is $\#\mathcal{A}^{\#\mathcal{A}^n}$, as the function values of all the $\#\mathcal{A}^n$ inputs can be chosen freely. Hence, we have:

Theorem 3.15. *(1) If the carrier of a Boolean algebra* $\mathcal{B} = (\mathcal{A}; +, \cdot, ^-, 0, 1)$ *consists of more than two elements, then there are functions* $\mathcal{A}^n \to \mathcal{A}$ *which are not Boolean functions.*

(2) If the carrier consists of exactly two elements, then each function $\mathcal{A}^n \to \mathcal{A}$ *is a Boolean function.* $\square$

This fundamental fact is one of the reasons why Boolean algebras with two elements, investigated in detail in Section 3.3, are of particular importance.

Example 3.16. Let $S = \{1,2\}$ and $(2^S; \cup, \cap, ^-, \emptyset, S)$ be the set algebra of S. The function $f : S^4 \to S$,

$$f(x_1, x_2, x_3, x_4) = \begin{cases} \{2\} & \text{if } (x_1, x_2, x_3, x_4) = (\emptyset, \emptyset, \emptyset, \emptyset), \\ \{1\} & \text{otherwise,} \end{cases}$$

is not a Boolean function. This can be seen as follows. In the notation of the previous proof we have

$$a(0, 0, 0, 0) = f(\emptyset, \emptyset, \emptyset, \emptyset) = \{2\},$$
$$a(1, 0, 0, 0) = f(S, \emptyset, \emptyset, \emptyset) = \{1\}.$$

Hence,

$$f(\{1\}, \emptyset, \emptyset, \emptyset) = \bigcup_{(e_1, \dots, e_4) \in \{0,1\}^4} a(e_1, e_2, e_3, e_4) \cap \{1\}^{e_1} \cap \emptyset^{e_2} \cap \emptyset^{e_3} \cap \emptyset^{e_4}$$
$$= (a(1, 0, 0, 0) \cap \{1\}) \cup (a(0, 0, 0, 0) \cap \{2\}) = \{1\} \cup \{2\} = S,$$

in contradiction to the definition of f.   ◇

3.3 Switching Functions

In the following, we discuss the special case of Boolean algebra with two elements, which can be seen as the theoretical foundation of circuit design. The significance of this connection was recognized very early. In 1910, the physician P. Ehrenfest vaguely suggested applying Boolean algebra to the design of switching systems (in telephone networks). Much more detailed treatments appeared independently between 1936 and 1938 in Japan, the United States, and the Soviet Union. Without doubt, the most influential paper in this period was C. E. Shannon's "Symbolic Analysis of Relay and Switching Circuits". Nowadays, the developed calculus of switching functions is widely used and applied.

Definition 3.17. *Let the operations $+, \cdot, ^-$ on the set $\{0, 1\}$ be defined as follows:*

- $a + b = \max\{a, b\}$,
- $a \cdot b = \min\{a, b\}$,
- $\bar{0} = 1, \bar{1} = 0$.

Then $(\{0, 1\}, +, \cdot, ^-, 0, 1)$ is a Boolean algebra, called the **switching algebra**. *In the following, the set $\{0, 1\}$ is denoted by $\mathbb{B}$, and the term $a \cdot b$ is abbreviated by ab.*

According to Theorem 3.7, all Boolean algebras with exactly two elements are isomorphic to the switching algebra.

Definition 3.18. *An n-variable function $f : \mathbb{B}^n \to \mathbb{B}$ is called a **switching function**. The set of all n-variable switching functions is denoted by $\mathbb{B}_n$.*

As a consequence of Theorem 3.15, each n-variable switching function is a Boolean function. If it is clear from the context that the underlying Boolean algebra is the switching algebra, the terms *Boolean function* and *switching function* can be used synonymously.

If C is a circuit with n inputs signals and m output signals, then the input/output behavior of C can be described in terms of an m-tuple $f = (f_1, \dots, f_m)$ of switching functions. The set of n-variable switching functions with m outputs is denoted by $\mathbb{B}_{n,m}$. Obviously, $\mathbb{B}_{n,1}$ coincides with the already defined set $\mathbb{B}_n$.

The immense complexity observed when dealing with switching functions is caused substantially by the very rapid growth of the cardinality of $\mathbb{B}_{n,m}$.

Theorem 3.19. *The number of different n-variable switching functions with m outputs amounts to*

$$\#\mathbb{B}_{n,m} = 2^{m2^n}.$$

In particular, the number of n-variable switching functions with one output amounts to

$$\#\mathbb{B}_n = 2^{2^n}.$$

Proof. Let A and B be finite sets. Then there are $\#B^{\#A}$ different mappings from A to B. For $A = \mathbb{B}^n$ and $B = \mathbb{B}^m$ we obtain $\#\mathbb{B}_{n,m} = (2^m)^{2^n} = 2^{m2^n}$. $\square$

The immense growth of $\#\mathbb{B}_{n,m}$ can be illustrated by the values in Fig. 3.3 which shows the cardinalities of $\mathbb{B}_n$ and $\mathbb{B}_{n,m}$ for some of the *smallest* values of n and m.

$$\begin{aligned}
\#\mathbb{B}_2 &= 2^4 = 16 & \#\mathbb{B}_{2,2} &= 2^8 & \#\mathbb{B}_{2,3} &= 2^{12} \\
\#\mathbb{B}_3 &= 2^8 = 256 & \#\mathbb{B}_{3,2} &= 2^{16} & \#\mathbb{B}_{3,3} &= 2^{24} \\
\#\mathbb{B}_4 &= 2^{16} = 65536 & \#\mathbb{B}_{4,2} &= 2^{32} & \#\mathbb{B}_{4,3} &= 2^{48} \\
\#\mathbb{B}_5 &= 2^{32} > 4 \cdot 10^9 & \#\mathbb{B}_{5,2} &= 2^{64} & \dots \\
\#\mathbb{B}_6 &= 2^{64} > 16 \cdot 10^{18} & \dots & \dots
\end{aligned}$$

Figure 3.3. Growth of the number of Boolean functions

Before going into some special classes of switching functions, we will introduce some basic notation.

Definition 3.20. *Let $f \in \mathbb{B}_n$.*

1. *A vector $a = (a_1, \ldots, a_n) \in \mathbb{B}^n$ is called a* **satisfying assignment** *of f if $f(a) = 1$.*
2. *The set of all satisfying assignments and the set of all non-satisfying assignments of f are called the* **on-set** *of f and* **off-set** *of f, respectively.*

$$\mathrm{on}(f) = \{(a_1, \ldots, a_n) \in \mathbb{B}^n : f(a_1, \ldots, a_n) = 1\},$$
$$\mathrm{off}(f) = \{(a_1, \ldots, a_n) \in \mathbb{B}^n : f(a_1, \ldots, a_n) = 0\}.$$

3. *A variable x_i is called* **essential** *for f if there exists an input assignment $(a_1, \ldots, a_n)$ with*

$$f(a_1, \ldots, a_n, 0, a_{i+1}, \ldots, a_n) \neq f(a_1, \ldots, a_n, 1, a_{i+1}, \ldots, a_n).$$

A switching function f can be identified with its on-set. This identification is justified, as its on-set uniquely describes the function f. More precisely, for each switching function $f \in \mathbb{B}_n$ we have the relation

$$f = \chi_{\mathrm{on}(f)},$$

where $\chi_{\mathrm{on}(f)}$ denotes the characteristic function of $\mathrm{on}(f)$, i.e.,

$$\chi_{\mathrm{on}(f)}(a) = \begin{cases} 1 & \text{if } a \in \mathrm{on}(f), \\ 0 & \text{otherwise.} \end{cases}$$

By using the notion of essential variables, two functions with different numbers of variables can be related to each other. For example, in order to consider functions from $\mathbb{B}_n$ as a subset of $\mathbb{B}_{n+1}$, a dummy variable has to be introduced. However, this dummy variable does not have any influence on the actual computation, and, hence, this variable is not essential in the sense of the above definition. As a consequence of this notion, two functions with different numbers of variables can be considered as equal if they are equal after eliminating all inessential variables.

3.3.1 Switching Functions with at Most Two Variables

Switching functions depending on at most two variables occur particularly often and therefore have their own terminology.

Switching functions in zero variables. The Boolean constants 0, 1 are the only two switching functions which do not possess any essential variables. These two functions, also known as **contradiction** and **tautology**, belong to any set $\mathbb{B}_n$, $n \geq 0$. For any input size n, they can be defined by

$$1(x_1, \ldots, x_n) = 1 \quad \text{and} \quad 0(x_1, \ldots, x_n) = 0.$$

Switching functions in one variable. According to Theorem 3.19, there are exactly four switching functions in a single variable x_1. Two of them are the constant functions $f_1(x_1) = 0$ and $f_2(x_1) = 1$.

A **literal** denotes a variable x_i or its complement $\overline{x_i}$. The two remaining switching functions in one variable are just the two literals, $f_3(x_1) = x_1$ and $f_4(x_1) = \overline{x_1}$.

Switching functions in two variables. Theorem 3.19 implies that there are 16 switching functions in two variables, which are also called **binary operations.** Among those, again, are the two constant functions as well as the four literals x_1, x_2, $\overline{x_1}$, $\overline{x_2}$. The remaining 10 functions of $\mathbb{B}_2$ have the property that both variables are essential. Due to their frequent application they are worth enumerating explicitly:

Disjunction (OR, $\vee$, +):
  $\text{OR}(x_1, x_2) = 1$ if and only if $x_1 = 1$ or $x_2 = 1$.
Conjunction (AND, $\wedge$, $\cdot$):
  $\text{AND}(x_1, x_2) = 1$ if and only if $x_1 = 1$ and $x_2 = 1$.
Equivalence (EQUIVALENCE, $\Leftrightarrow$, $\equiv$, =):
  $\text{EQUIVALENCE}(x_1, x_2) = 1$ if and only if $x_1 = x_2$.
Implication (IMPLY, $\Rightarrow$):
  $\text{IMPLY}(x_1, x_2) = 1$ if and only if $x_1 = 0$ or $x_1 = x_2 = 1$.
R-L-implication (IMPLIED-BY, $\Leftarrow$):
  $\text{IMPLIED-BY}(x_1, x_2) = 1$ if and only if $x_2 = 0$ or $x_1 = x_2 = 1$.

The other 5 functions can be obtained from these function by negation:

Negated disjunction (NOR, $\overline{+}$, $\overline{\vee}$):
  $\text{NOR}(x_1, x_2) = 1$ if and only if $x_1 = 0$ and $x_2 = 0$.
Negated conjunction (NAND, $\overline{\cdot}$, $\overline{\wedge}$):
  $\text{NAND}(x_1, x_2) = 1$ if and only if $x_1 = 0$ or $x_2 = 0$.
Parity function (EX-OR, EXOR, XOR, $\oplus$, mod2):
  $\text{EX-OR}(x_1, x_2) = 1$ if and only if $x_1 \neq x_2$.
Negated implication (NOT-IMPLY, $\not\Rightarrow$):
  $\text{NOT-IMPLY}(x_1, x_2) = 1$ if and only if $x_1 = 1$ and $x_2 = 0$.
Negated R-L-implication (NOT-IMPLIED-BY, $\not\Leftarrow$):
  $\text{NOT-IMPLIED-BY}(x_1, x_2) = 1$ if and only if $x_1 = 0$ and $x_2 = 1$.

Instead of the functional notation we will mostly prefer to use the more compact operator notation. We write

$$
\begin{array}{lll}
x_1 + x_2 & \text{instead of} & \text{OR}(x_1, x_2), \\
x_1 \cdot x_2 & \text{instead of} & \text{AND}(x_1, x_2), \\
x_1 \equiv x_2 & \text{instead of} & \text{EQUIVALENCE}(x_1, x_2), \\
x_1 \Rightarrow x_2 & \text{instead of} & \text{IMPLY}(x_1, x_2), \\
x_1 \Leftarrow x_2 & \text{instead of} & \text{IMPLIED-BY}(x_1, x_2), \\
x_1 \overline{+} x_2 & \text{instead of} & \text{NOR}(x_1, x_2), \\
x_1 \overline{\cdot} x_2 & \text{instead of} & \text{NAND}(x_1, x_2), \\
x_1 \oplus x_2 & \text{instead of} & \text{EX-OR}(x_1, x_2), \\
x_1 \not\Rightarrow x_2 & \text{instead of} & \text{NOT-IMPLY}(x_1, x_2), \\
x_1 \not\Leftarrow x_2 & \text{instead of} & \text{NOT-IMPLIED-BY}(x_1, x_2).
\end{array}
$$

The small size of the domain of two-variable switching functions allows us to present the functions $f \in \mathbb{B}_2$ in tabular form as in Fig. 3.4.

x_1	x_2	f_0	f_1	f_2	f_3	f_4	f_5	f_6	f_7	f_8	f_9	f_{10}	f_{11}	f_{12}	f_{13}	f_{14}	f_{15}
		0	$\cdot$	$\not\Rightarrow$	x_1	$\not\Leftarrow$	x_2	$\oplus$	$+$	$\overline{+}$	$\equiv$	$\overline{x_2}$	$\Leftarrow$	$\overline{x_1}$	$\Rightarrow$	$\overline{\cdot}$	1
0	0	0	0	0	0	0	0	0	0	1	1	1	1	1	1	1	1
0	1	0	0	0	0	1	1	1	1	0	0	0	0	1	1	1	1
1	0	0	0	1	1	0	0	1	1	0	0	1	1	0	0	1	1
1	1	0	1	0	1	0	1	0	1	0	1	0	1	0	1	0	1

Figure 3.4. Enumeration of binary switching functions

3.3.2 Subfunctions and Shannon's Expansion

An important foundation for computations with Boolean and switching function is the so-called Shannon expansion, first formulated by Boole. This theorem establishes an important link between a function and its subfunctions. Here, a function g is called **subfunction** of f if it results from f by assigning constants to some of its input variables. In many contexts, knowledge of the properties of certain subfunctions yields essential information about the function itself.

Theorem 3.21. (Shannon's expansion)
Let $f \in \mathbb{B}_n$ be an n-variable switching function. For the subfunctions $g, h \in \mathbb{B}_{n-1}$ defined by

$$g(x_1, \ldots, x_{n-1}) = f(x_1, \ldots, x_{n-1}, 0),$$
$$h(x_1, \ldots, x_{n-1}) = f(x_1, \ldots, x_{n-1}, 1)$$

it holds that

$$f = \overline{x_n} \cdot g + x_n \cdot h. \tag{3.1}$$

Proof. Let $a = (a_1, \ldots, a_n)$ be an assignment to the input variables $x_1, \ldots, x_n$. In case $a_n = 0$ the theorem immediately follows from the equation $f(a_1, \ldots, a_n) = g(a_1, \ldots, a_{n-1})$. In case $a_n = 1$ it follows from the equation $f(a_1, \ldots, a_n) = h(a_1, \ldots, a_{n-1})$.   □

A subfunction originates from fixing one or several input variables. For this reason, each subfunction of a function $f \in \mathbb{B}_n$ can be specified by a vector $c \in \{0, 1, \mathrm{id}\}^n$. If the i-th component of c is the constant 0 or 1, the i-th input variable x_i in f is set to 0 or 1, respectively; if c_i has the value id, then the variable x_i remains unfixed. Hence, the vector c defines the subfunction

$$f_c(x_1, \ldots, x_n) := f(c_1(x_1), \ldots, c_n(x_n)).$$

From this definition of a subfunction, it follows immediately that a switching function can have at most 3^n subfunctions.

Corollary 3.22. *An n-variable switching function $f \in \mathbb{B}_n$ has at most 3^n different subfunctions.*   □

In general, the subfunctions of a function f are not pairwise different.

Example 3.23. The switching function $f(x_1, x_2, x_3) = x_1 + (x_2 \cdot \overline{x_3})$ has 9 different subfunctions:

$f_c = 0$	for	$c \in \{(0,0,0), (0,0,1), (0,0,\mathrm{id}), (0,1,1),$
		$(0,\mathrm{id},1)\}$,
$f_c = 1$	for	$c \in \{(0,1,0), (1,0,0), (1,0,1), (1,0,\mathrm{id}),$
		$(1,1,0), (1,1,1), (1,1,\mathrm{id}), (1,\mathrm{id},0),$
		$(1,\mathrm{id},1), (1,\mathrm{id},\mathrm{id}), (\mathrm{id},1,0)\}$,
$f_c = x_1$	for	$c \in \{(\mathrm{id},0,0), (\mathrm{id},0,1), (\mathrm{id},0,\mathrm{id}), (\mathrm{id},1,1),$
		$(\mathrm{id},\mathrm{id},1)\}$,
$f_c = x_2$	for	$c \in \{(0,\mathrm{id},0)\}$,
$f_c = \overline{x_3}$	for	$c \in \{(0,1,\mathrm{id})\}$,
$f_c = x_1 + x_2$	for	$c \in \{(\mathrm{id},\mathrm{id},0)\}$,
$f_c = x_1 + \overline{x_3}$	for	$c \in \{(\mathrm{id},1,\mathrm{id})\}$,
$f_c = x_2\overline{x_3}$	for	$c \in \{(0,\mathrm{id},\mathrm{id})\}$,
$f_c = x_1 + x_2\overline{x_3}$	for	$c \in \{(\mathrm{id},\mathrm{id},\mathrm{id})\}$.

◇

In Shannon's expansion of Equation (3.1), the involved subfunctions g and h are defined by the vectors $c = (\mathrm{id}, \ldots, \mathrm{id}, 0)$ and $c' = (\mathrm{id}, \ldots, \mathrm{id}, 1)$. Of course, the idea of Shannon's expansion can also be transferred to other subfunctions and to other operations.

Corollary 3.24. *For each function $f \in \mathbb{B}_n$ and each $1 \le i \le n$ the following holds:*

1. *Shannon's expansion of the i-th argument:*

$$
\begin{aligned}
f(x_1, \ldots, x_n) = {}& x_i\ f(x_1, \ldots, x_{i-1}, 1, x_{i+1}, \ldots, x_n) \\
& + \ \overline{x_i}\ f(x_1, \ldots, x_{i-1}, 0, x_{i+1}, \ldots, x_n).
\end{aligned}
$$

2. *Dual of Shannon's expansion:*

$$
\begin{aligned}
f(x_1, \ldots, x_n) = {}& (x_i\ + \ f(x_1, \ldots, x_{i-1}, 0, x_{i+1}, \ldots, x_n)) \\
& \cdot \ (\overline{x_i}\ + \ f(x_1, \ldots, x_{i-1}, 1, x_{i+1}, \ldots, x_n)).
\end{aligned}
$$

3. *Shannon's expansion with respect to $\oplus$:*

$$
\begin{aligned}
f(x_1, \ldots, x_n) = {}& x_i\ f(x_1, \ldots, x_{i-1}, 1, x_{i+1}, \ldots, x_n) \\
& \oplus\ \overline{x_i}\ f(x_1, \ldots, x_{i-1}, 0, x_{i+1}, \ldots, x_n).
\end{aligned}
$$

$\square$

Proof. The first statement is clear. The dual of Shannon's expansion is a consequence of the principle of duality. Shannon's expansion with respect to $\oplus$ follows from the fact that for any assignment to x_i either the first or the second of the $\oplus$-terms vanishes. $\square$

3.3.3 Visual Representation

To make use of geometric imagination and intuition in investigating Boolean functions, the following geometric illustration of the domain $\mathbb{B}^n = \{0, 1\}^n$ of switching functions has turned out to be very fruitful. The single arguments $a = (a_1, \ldots, a_n) \in \mathbb{B}^n$ are interpreted as vertex coordinates of an n-dimensional unit cube, so one can imagine $\mathbb{B}^n$ as embedded in the n-dimensional Euclidean space $\mathbb{R}^n$ (see Fig. 3.5). Due to this form of visualization the set $\mathbb{B}^n$ is often called the n-dimensional **Boolean cube**.

Although it becomes quite hard to imagine a 4-,7-, or a 37-dimensional cube, analogies from low-dimensional spaces can be used to deduce information about the relations in higher dimensions. Maybe the representation of a 4-dimensional unit cube in Fig. 3.6 gives an initial idea in this direction.

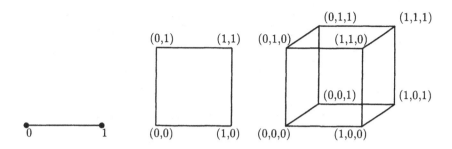

Figure 3.5. The one-, two-, and three-dimensional Boolean cubes with the coordinates of their vertices

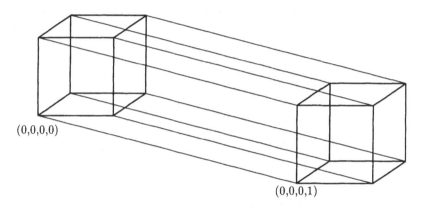

Figure 3.6. The 4-dimensional Boolean cube

As a switching function $f \in \mathbb{B}_n$ assigns to each argument $a \in \mathbb{B}^n$ one of only two possible values, $f(a) \in \{0, 1\}$, the cube visualizing the domain can also be used to geometrically represent the function itself. For example, exactly those vertices of the cube which represent an element in the on-set on(f) are marked.

Example 3.25. Figure 3.7 shows a tabular representation and the geometric visualization of the switching function $f(x_1, x_2, x_3) = x_1 + (x_2 \cdot \overline{x_3})$.   ◇

3.3.4 Monotone Switching Functions

In general, the analysis of switching functions with many input variables is very difficult, as the treatment may require a large amount of resources (e.g.,

x_1	x_2	x_3	$f(x_1, x_2, x_3)$
0	0	0	0
0	0	1	0
0	1	0	1
0	1	1	0
1	0	0	1
1	0	1	1
1	1	0	1
1	1	1	1

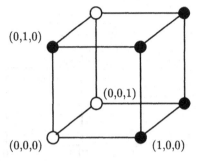

Figure 3.7. Tabular representation and geometric visualization

time, space) that can obviously not be provided. However, under certain circumstances, this situation can change dramatically if additional structural properties of the occurring functions are known. One important property in this context is **monotony**.

Definition 3.26. *Let $f \in \mathbb{B}_n$, and let $1 \leq i \leq n$. f is called* **monotone increasing** *in the i-th input if for all $a \in \mathbb{B}^n$:*

$$f(a_1, \ldots, a_{i-1}, 0, a_{i+1}, \ldots, a_n) \leq f(a_1, \ldots, a_{i-1}, 1, a_{i+1}, \ldots, a_n).$$

f is called **monotone decreasing** *in the i-th input if for all $a \in \mathbb{B}^n$:*

$$f(a_1, \ldots, a_{i-1}, 1, a_{i+1}, \ldots, a_n) \leq f(a_1, \ldots, a_{i-1}, 0, a_{i+1}, \ldots, a_n).$$

If f is monotone increasing (respectively monotone decreasing) in each argument, then f is called **monotone increasing** *(respectively* **monotone decreasing***).*

The following characterization of monotone increasing and monotone decreasing functions could be equivalently used for defining monotony.

Theorem 3.27. *(1) A function $f \in \mathbb{B}_n$ is monotone increasing if and only if for all $a, b \in \mathbb{B}^n$ the property $a \leq b$ implies $f(a) \leq f(b)$.*
(2) A function $f \in \mathbb{B}_n$ is monotone decreasing if and only if for all $a, b \in \mathbb{B}^n$ the property $a \leq b$ implies $f(a) \geq f(b)$.

Proof. Statements (1) and (2) can be proven by arguments that are completely analogous to each other. Hence, we only consider the first statement. Let f be monotone increasing. Then, by definition, f is monotone increasing in each argument. Now let $a = (a_1, \ldots, a_n)$, $b = (b_1, \ldots, b_n) \in \mathbb{B}^n$ with $a \leq b$, and let $a_i < b_i$ for $i = i_1, \ldots, i_k$. If in the vector a the components a_{i_j} are successively replaced by the components b_{i_j}, $1 \leq j \leq k$, then, due to monotony in the i_j-th argument, one obtains a chain of inequalities:

$$f(a) \leq \ldots \leq f(b).$$

This proves $f(a) \leq f(b)$.

If, conversely, the property $a \leq b$ implies $f(a) \leq f(b)$, then in particular, for all i, $1 \leq i \leq n$ and for all $a \in \mathbb{B}^n$ we have

$$f(a_1, \ldots, a_{i-1}, 0, a_{i+1}, \ldots, a_n) \leq f(a_1, \ldots, a_{i-1}, 1, a_{i+1}, \ldots, a_n).$$

Hence, f is monotone increasing in each argument.   □

Example 3.28. (1) Obviously, the literal x_i, $1 \leq i \leq n$, is monotone increasing in the i-th argument. In all other arguments, the literal is both monotone increasing and monotone decreasing. Altogether, each literal x_i is a monotone increasing function. Analogously, the literals $\overline{x_i}$ are monotone decreasing functions.

(2) The following two functions

$$f(x_1, \ldots, x_n) = x_{i_1} \cdot \ldots \cdot x_{i_k}, \quad 1 \leq i_1 \leq \ldots \leq i_k \leq n,$$
$$g(x_1, \ldots, x_n) = x_{i_1} + \ldots + x_{i_k}, \quad 1 \leq i_1 \leq \ldots \leq i_k \leq n$$

are monotone increasing in each argument. Moreover, in the inessential arguments, f and g are also monotone decreasing.   ◇

Obviously, in all inessential arguments, Boolean functions are both monotone increasing and monotone decreasing. More interesting is the question of monotony in the essential variables. An important contribution to this question is provided by the following **representation theorem for monotone functions in the i-th argument** which we state without proof.

Theorem 3.29. *(1) A switching function $f \in \mathbb{B}_n$ is monotone increasing in the i-th argument if and only if f can be represented in the form*

$$f \;=\; x_i \, g \,+\, h$$

in terms of two functions g and h which are independent of x_i.

(2) A switching function $f \in \mathbb{B}_n$ is monotone decreasing in the i-th argument if and only if f can be represented in the form

$$f \;=\; \overline{x_i} \, g \,+\, h$$

in terms of two functions g and h which are independent of x_i.

Monotone functions occur in many fundamental applications, such as sorting or matrix multiplication.

Example 3.30. (Matrix multiplication)
Let $X = (x_{ij}), Y = (y_{ij}) \in \mathbb{B}^{n,n}$ be two $n \times n$-matrices. The matrix product $Z := X \cdot Y$ is defined by the elements

$$z_{ij} = \sum_{k=1}^{n} x_{ik} y_{kj}.$$

For all $1 \leq i, j \leq n$ the functions z_{ij} are monotone increasing.   ◇

3.3.5 Symmetric Functions

Similar to the case of monotony, symmetry properties can also be exploited in order to make manipulation of switching function much easier. A variety of algorithms whose application to arbitrary functions is not practical work efficiently in case of symmetric functions.

Definition 3.31. *A switching function $f \in \mathbb{B}_n$ is called* **symmetric** *if each permutation π of the input variables does not change the function value, i.e.,*

$$f(x_1, \dots, x_n) = f(x_{\pi(1)}, \dots, x_{\pi(n)}).$$

Obviously, a function f is symmetric if and only if its function value only depends on the number of 1's in the input vector and not on their positions. Hence, a symmetric function $f \in \mathbb{B}_n$ can be represented by a vector $v(f) = (v_0, \dots, v_n) \in \mathbb{B}^{n+1}$ of length $n + 1$ in a natural way. The components $v_0, \dots, v_n$ are defined by

$$v_i := f(\underbrace{1, \dots, 1}_{i \text{ times}}, 0, \dots, 0)$$

and represent the function value of f if exactly i input bits are 1 and $n-i$ input bits are 0. $v(f)$ is called the **spectrum** or **value vector** of f. The one-to-one-correspondence between symmetric functions and the 2^{n+1} possible value vectors immediately implies that there are exactly 2^{n+1} symmetric functions in n variables.

Corollary 3.32. *There are 2^{n+1} symmetric functions in n variables.*   □

Example 3.33. The following functions are symmetric:

$$f(x_1, x_2, x_3) = x_1 \oplus x_2 \oplus x_3,$$
$$g(x_1, x_2, x_3, x_4) = x_1 + x_2 + x_3 + x_4,$$
$$h(x_1, x_2, x_3) = x_1 x_2 + x_2 x_3 + x_1 x_3.$$

The corresponding value vectors are

$$v(f) = (0, 1, 0, 1),$$
$$v(g) = (0, 1, 1, 1, 1),$$
$$v(h) = (0, 0, 1, 1).$$

In contrast to this, the following functions are not symmetric:

$$f(x_1, x_2, x_3) = x_1 x_2 + x_3,$$
$$g(x_1, x_2, x_3) = \overline{x_1} x_2 + x_1 x_3,$$
$$h_i(x_1, \dots, x_n) = x_i, \quad 1 \le i \le n, \quad n \ge 2.$$

$\diamond$

As the number of symmetric functions is very small in comparison to the total number 2^{2^n} of all Boolean function in n variables, a randomly chosen function is symmetric only with a very small probability. However, functions which are important for practical applications are not randomly chosen at all. In fact, symmetric functions appear very often, e.g., when the underlying problem is related to a counting problem. The following functions are of particular interest – some of them already appeared in Example 3.33.

Parity function $\text{PAR}_n \in \mathbb{B}_n$:

$$\text{PAR}_n(x_1, \dots, x_n) = \oplus_{i=1}^{n} x_i = \left(\sum_{i=1}^{n} x_i \right) \pmod 2,$$

Majority function $M_n \in \mathbb{B}_n$:

$$M_n(x_1, \dots, x_n) = 1 \text{ if and only if } \sum_{i=1}^{n} x_i \ge \frac{n}{2},$$

Threshold functions $T_k^n \in \mathbb{B}_n$, $0 \le k \le n$:

$$T_k^n(x_1, \dots, x_n) = 1 \text{ if and only if } \sum_{i=1}^{n} x_i \ge k,$$

Inverse threshold functions $T_{\le k}^n \in \mathbb{B}_n$, $0 \le k \le n$:

$$T_{\le k}^n(x_1, \dots, x_n) = 1 \text{ if and only if } \sum_{i=1}^{n} x_i \le k,$$

Interval functions $I_{k,l}^n \in \mathbb{B}_n$, $1 \le k \le l \le n$:

$$I_{k,l}(x_1, \dots, x_n) = 1 \text{ if and only if } k \le \sum_{i=1}^{n} x_i \le l.$$

Example 3.34. Some relations among important classes of switching functions:

$$x_1 + x_2 + \ldots + x_n = I_{1,n}^n(x_1, \ldots, x_n),$$
$$x_1 \cdot x_2 \cdot \ldots \cdot x_n = I_{n,n}^n(x_1, \ldots, x_n),$$
$$M_n(x_1, \ldots, x_n) = T_{\lceil \frac{n}{2} \rceil, n}^n(x_1, \ldots, x_n),$$
$$T_{\leq k}^n = I_{0,k}^n(x_1, \ldots, x_n).$$

$\diamond$

The possibility of representing symmetric functions in terms of interval functions is not exhausted by the examples that have just been stated. In fact, every symmetric function can be represented by a disjunction of interval functions. This statement follows from the fact that the value vectors $v(f)$ and $v(g)$ of two symmetric functions f and g satisfy the following equation:

$$v(f \vee g) = v(f) \vee v(g).$$

Example 3.35. Let $f \in \mathbb{B}_9$ be a symmetric function with spectrum $v(f) = (1, 0, 0, 1, 0, 0, 1, 1, 1, 0)$. The spectrum can be written in the form

$$v(f) = v(I_{0,0}^9) \vee v(I_{3,3}^9) \vee v(I_{6,8}^9).$$

Hence, f can be represented by the disjunction of interval functions,

$$f = I_{0,0}^9 \vee I_{3,3}^9 \vee I_{6,8}^9.$$

$\diamond$

Symmetry effects may sometimes also be used in case of functions which are not "totally" symmetric. Two restricted types of symmetry are presented in the following definition.

Definition 3.36. *(1)   A function $f(x_1, \ldots, x_n) \in \mathbb{B}_n$ is called* **quasisymmetric** *if substituting some variables by their complement leads to a symmetric function.*

(2)   A function $f(x_1, \ldots, x_n) \in \mathbb{B}_n$ is called **partially symmetric** *with respect to the variables $x_{i_1}, \ldots, x_{i_k}$ if f remains invariant for all permutations of the variables $x_{i_1}, \ldots, x_{i_k}$.*

Example 3.37. The following functions $f, g \in \mathbb{B}_3$ are quasisymmetric:

$$f(x_1, x_2, x_3) = x_1 \overline{x_2} x_3,$$
$$g(x_1, x_2, x_3) = \overline{x_1} x_2 + x_2 \overline{x_3} + \overline{x_1}\, \overline{x_3}.$$

Moreover, both f and g are partially symmetric with respect to the variables x_1, x_3.

$\diamond$

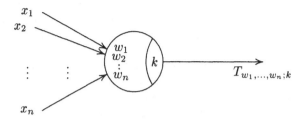

Figure 3.8. Threshold function

3.3.6 Threshold Functions

Within the presentation of symmetric switching functions we have already mentioned the class of threshold functions T_k^n. This class is of great importance for modeling biological neurons and therefore for constructing artificial neural networks. In the biological scenario, the neuron's cell membrane is capable of sustaining a certain electric charge. When this charge reaches or exceeds a threshold k, the neuron "fires." In the terminology of switching functions, this effect is modeled by means of threshold functions. Threshold functions are not only symmetric but also monotone increasing.

In order to cover different input resistances in the biological setting, the following notation, illustrated in Fig. 3.8, is quite helpful.

Definition 3.38. *A switching function* $f \in \mathbb{B}_n$ *is called a* **weighted threshold function with weights** $w_1, \ldots, w_n \in \mathbb{R}$ *and threshold* $k \in \mathbb{R}$ *if*

$$f(x_1, \ldots, x_n) = 1 \quad \text{if and only if} \quad \sum_{i=1}^{n} w_i x_i \geq k.$$

We write $T_{w_1, \ldots, w_n; k}$.

Although the large number of parameter choices makes this class very powerful in its expressiveness, there are nevertheless switching functions which cannot be represented in terms of a weighted threshold function.

Example 3.39. Let us consider the function $f(x_1, x_2) = x_1 \overline{x_2} + x_2 \overline{x_1}$. This function cannot be represented in terms of a weighted threshold function. This can be explained as follows. If there were real weights w_1, w_2 and a threshold value k satisfying $f = T_{w_1, w_2; k}$, then the following inequalities would hold:

$$0 \cdot w_1 + 1 \cdot w_2 \geq k \quad (\text{since } f(0,1) = 1),$$
$$1 \cdot w_1 + 0 \cdot w_2 \geq k \quad (\text{since } f(1,0) = 1),$$
$$1 \cdot w_1 + 1 \cdot w_2 < k \quad (\text{since } f(1,1) = 0).$$

The first two inequalities can be combined to $w_1 + w_2 \geq 2k$, which contradicts the third equation. Consequently, the function f cannot be represented in terms of a weighted threshold function.   ◇

An important structural property of weighted threshold functions says that in each input the function is monotone increasing or monotone decreasing. A proof of this statement can be carried out by generalizing the arguments in the previous example.

3.3.7 Partial Switching Functions

When describing the functional behavior of a digital system, it is often not necessary to specify this behavior for absolutely all possible input constellations. In many cases it is known a priori that certain constellations do not occur, for example if the input of the system results from the output of another one, or if the reaction of the system on certain constellations has no noticeable effect. Hence, appropriate functions that model the system mathematically only need to be partially defined. A **partial switching function** in n inputs is a mapping from $\mathbb{B}^n$ to $\mathbb{B}$ which is merely defined on a subset of $\mathbb{B}^n$, namely on the set $\mathrm{def}(f) \subset \mathbb{B}^n$. For the set of all partial functions from $\mathbb{B}^n$ to $\mathbb{B}$ we write $\mathbb{B}_n^*$.

If, as in the case of completely specified functions, $\mathrm{on}(f)$ and $\mathrm{off}(f)$ denote the set of input vectors of f mapping to 0 and 1, respectively, then $\mathrm{def}(f) = \mathrm{on}(f) \cup \mathrm{off}(f)$. The set $\mathrm{dc}(f) = \mathbb{B}^n - \mathrm{def}(f)$ of all inputs for which f remains undefined is called the **don't care set**. Each partial switching function is completely described by at least two of the three sets $\mathrm{on}(f)$, $\mathrm{off}(f)$, and $\mathrm{dc}(f)$.

Example 3.40. Let $f \in \mathbb{B}_3^*$ be a partial switching function defined on the four inputs $(0,0,1),(0,1,0),(1,1,0),(1,1,1)$ by

$$f(0,0,1) = f(0,1,0) = f(1,1,1) = 1, \text{ and } f(1,1,0) = 0.$$

Then the don't care set consists of the inputs $(0,0,0)$, $(0,1,1)$, $(1,0,0)$, $(1,0,1)$.   ◇

By choosing arbitrary function values for the inputs of the don't care set, it is possible to completely specify a partial switching function $f \in \mathbb{B}_n^*$. We say that the function f is **embedded** in the set $\mathbb{B}_n$ of all completely specified functions. The resulting function $f' \in \mathbb{B}_n$ is called an **extension of** f. Obviously, the number of possible extensions of a function $f \in \mathbb{B}_n^*$ is $2^{\#\mathrm{dc}(f)}$. Keeping in mind that digital circuits only realize completely defined switching functions, one recognizes the significance of the task of constructing best suitable extensions with respect to given optimization criteria.

Example 3.41. The function f from Example 3.40 has 2^4 possible extensions $f' \in \mathbb{B}_n$. These extensions include the parity function $x_1 \oplus x_2 \oplus x_3$ and the interval function $I_{1,3}$.   ◇

An n-variable **partial switching function** f **with** m **outputs** is a tuple of m partially defined switching functions $f_i \in \mathbb{B}_n^*$, $1 \leq i \leq m$,

$$f = (f_1, \ldots, f_m).$$

3.4 References

The material in this chapter belongs to the "classical" knowledge of Boolean algebra and Boolean functions. An extensive presentation can be found, e.g., in [Bro90]. The mentioned historic books and papers by Boole, Shannon, and Ehrenfest are [Boo54, Sha38, Ehr10].

Furthermore, we refer the interested reader to the monograph [Weg87], which contains a comprehensive presentation of the complexity of Boolean functions.

4. Classical Representations

A classic is something that everybody wants to have read and nobody wants to read.

Mark Twain (1835–1910)

To be able to work with concrete switching functions, a description of these functions is required. Of course, there is a very broad spectrum of possible description types. Basically, the representation has to describe the function adequately and thoroughly, i.e., it must be completely clear from the representation which switching function is considered. Besides this fundamental condition, which has always to be satisfied, a series of further properties are desirable. For example, the description of a function should

- be rather short and efficient,
- support the evaluation and manipulation of the function,
- make particular properties of the function visible,
- suggest ideas for a technical realization,

and much more.

When looking for representation types which satisfy all the above properties, one very soon faces inherent difficulties. These difficulties are caused by the (even for small values of n) inconceivably large number of 2^{m2^n} switching functions with n inputs and m outputs. If only a small number of functions were of interest, one could construct a catalog in which all these functions were listed together with an enumeration of their properties. In this case, representing individual functions could be achieved by using the corresponding numbers of the entries within the catalog. Unfortunately, the large number of switching functions necessarily leads to numbers in the order 2^{m2^n}, and hence to representations of exponential length $m2^n$.

In this chapter, we present classical representation types which, thanks to their properties, have become important in theoretical investigations and practical applications within CAD systems.

However, all these representations have also some drawbacks. With permanently increasing performance requirements, these negative properties have become much more significant. For this reason, we also discuss the drawbacks of the individual representations.

4.1 Truth Tables

As the set $\mathbb{B}^n$ is finite, each switching function – at least in principle – allows a complete tabular listing of its arguments together with the corresponding function values. In fact, this representation in terms of **truth tables** is of great practical importance for small n (e.g., $n \leq 7$). Incidentally, truth tables were already used in the previous chapter, for instance in Example 3.25.

Example 4.1. Figure 4.1 shows the truth table of the multiplication function $MUL_{2,2} \in \mathbb{B}_{4,4}$, a function with four outputs. Here, $MUL_{2,2} = MUL(x_1, x_0, y_1, y_0)$ is a switching function interpreting the inputs $x_1 x_0$ and $y_1 y_0$ as binary numbers of length 2, and computing the binary representation of their product.   ◇

x_1 x_0 y_1 y_0	$(MUL_{2,2})_3$	$(MUL_{2,2})_2$	$(MUL_{2,2})_1$	$(MUL_{2,2})_0$
0 0 0 0	0	0	0	0
0 0 0 1	0	0	0	0
0 0 1 0	0	0	0	0
0 0 1 1	0	0	0	0
0 1 0 0	0	0	0	0
0 1 0 1	0	0	0	1
0 1 1 0	0	0	1	0
0 1 1 1	0	0	1	1
1 0 0 0	0	0	0	0
1 0 0 1	0	0	1	0
1 0 1 0	0	1	0	0
1 0 1 1	0	1	1	0
1 1 0 0	0	0	0	0
1 1 0 1	0	0	1	1
1 1 1 0	0	1	1	0
1 1 1 1	1	0	0	1

Figure 4.1. Truth table of the multiplication function $MUL_{2,2}$

The advantage of truth tables is their easy algorithmic handling. If, for example, the tables of two functions $f, g \in \mathbb{B}_n$ are given, then the functions can be evaluated very easily, or easily combined by means of binary operators. With regard to classical complexity theory which measures complexity in terms of the input length, only linear time resources are required. However, this statement completely disguises the fact that a truth table of an n-variable switching function consists of 2^n rows, i.e., the size of the table and hence the mentioned linear time complexity are always exponential in the number n of primary inputs. The size of the truth table of a partial switching function $f \in \mathbb{B}_n^*$ is proportional to the size of the definition domain $\operatorname{def}(f)$ and hence typically also exponential in n. Consequently, representations of switching functions in terms of truth tables are far from being compact.

4.2 Two-Level Normal Forms

Switching functions can be represented in terms of expressions that combine literals by means of suitable operators. Here, normal forms based on levelized expressions are of particular importance. The number of **levels** denotes the number of iterated applications of operators, where within a level, only one and the same operator may be employed.

One-level normal forms merely use a single operator. An easy counting argument immediately explains that the expressiveness of one-level normal forms is quite limited. Only very few switching functions can be represented in terms of these normal forms, e.g., the sum of literals or their product.

Two-level normal forms are significantly more powerful in their expressiveness. The following considerations show that all switching functions can be represented in terms of these normal forms.

Most work in the early days of circuit theory was dedicated to the investigation of two-level normal forms. This was mainly due to the fact that two-level forms can immediately be technically realized, e.g., by using programmable logic arrays (PLAs).

Definition 4.2. *Let $\omega = *_\omega \in \mathbb{B}_2$ be an associative Boolean operation. An ω-monomial m is an ω-product of literals*

$$m = x_{i_1}^{e_{i_1}} *_\omega x_{i_2}^{e_{i_2}} *_\omega \ldots *_\omega x_{i_l}^{e_{i_l}},$$

*$e_{i_j} \in \{0,1\}$, $j = 1, \ldots, l$. Here, x_i^0 and x_i^1 are defined by $x_i^1 = x_i$ and $x_i^0 = \overline{x_i}$. The number l of occurring literals is called the **length** of m.*

Example 4.3.

$\cdot$ **-monomials:** $x_1^0\, x_2\, x_3^0, x_1^0\, x_4,\; x_1\, x_2\, \overline{x_5}\, \overline{x_7}\, x_8,\; \ldots$

$+$ **-monomials:** $x_1^0 + x_2 + x_3^0 + x_1^0 + x_4,\; x_1 + x_2 + \overline{x_5} + \overline{x_7} + x_8,\; \ldots$

$\oplus$ **-monomials:** $x_1^0 \oplus x_2 \oplus x_3^0 \oplus x_1^0 \oplus x_4,\; x_1 \oplus x_2 \oplus \overline{x_5} \oplus \overline{x_7} \oplus x_8,\; \ldots$ $\diamond$

$\cdot$ -monomials $m = x_{i_1}^{e_{i_1}} \ldots x_{i_l}^{e_{i_l}}$ are simply called **monomials**. As already mentioned, the operator $\cdot$ is often omitted. The empty monomial is defined to be the tautology function 1. $+$ -monomials $m = x_{i_1}^{e_{i_1}} + \ldots + x_{i_l}^{e_{i_l}}$ are called **clauses**. The empty clause is defined to be the contradiction function 0.

Definition 4.4. *Let* $\omega = *_\omega$, $\omega' = *_{\omega'} \in \mathbb{B}_2$ *be two associative Boolean operations. An* (ω, ω')-**polynomial** *is an* ω'-*product of* ω-*monomials. The length of an* (ω, ω')-*polynomial is the sum of lengths of its monomials.*

$(\cdot, +)$-**polynomials** *are called* **disjunctive normal forms** *(DNF).*

$(+, \cdot)$-**polynomials** *are called* **conjunctive normal forms** *(CNF).*

$(\cdot, \oplus)$-**polynomials** *are called* **parity normal forms** *(PNF).*

We assume that there are no trivial redundancies like multiple occurrences of literals or terms, and that the order of terms is of not particular importance.

Example 4.5.

DNF representation: $d = x_1 \overline{x_0} y_1 + x_1 x_0 y_1 \overline{y_0}$,

CNF representation: $c = x_1 y_1 (\overline{x_1} + \overline{x_0} + \overline{y_1} + \overline{y_0})$,

PNF representation: $p = x_1 \overline{x_0} y_1 \oplus x_1 x_0 y_1 \overline{y_0}$. $\diamond$

Usually, one is interested in rather *short* representations.

Definition 4.6. *An* (ω, ω')-*polynomial of* $f \in \mathbb{B}_n$ *is called* **minimal**, *if its length is minimal among all* (ω, ω')-*polynomials of* f.

In order to generate (ω, ω')-polynomials from tabular representations, the following notations are useful. For $a = (a_1, \ldots, a_n) \in \mathbb{B}^n$, the monomial

$$m_a(x) = x_1^{a_1} \ldots x_n^{a_n}$$

is called the **minterm** of a, and

$$s_a(x) = x_1^{\overline{a_1}} + \ldots + x_n^{\overline{a_n}}$$

is called the **maxterm** of a. The following lemma turns out to be an immediate consequence of this notation.

Lemma 4.7. *For* $a = (a_1, \ldots, a_n) \in \mathbb{B}^n$ *we have:*

1. $m_a(x) = 1$ *if and only if* $x_i = a_i$ *for all* $1 \leq i \leq n$.
2. $s_a(x) = 0$ *if and only if* $x_i = a_i$ *for all* $1 \leq i \leq n$.   □

Due to this property, each function $f \in \mathbb{B}_n$ satisfies the equation

$$f(x) = \sum_{a \in \mathrm{on}(f)} m_a(x).$$

Indeed, this representation of f is unique. It is called the **canonical disjunctive normal form** (cDNF) of f. Analogously, f can be expressed in terms of

$$f(x) = \prod_{a \in \mathrm{off}(f)} s_a(x),$$

a representation which is called the **canonical conjunctive normal form** (cCNF) of f. The two canonical normal forms can be immediately deduced from a truth table.

Example 4.8. Using the notation of Fig. 4.1 we have:

$$(MUL_{2,2})_1 = x_1^0 \, x_0 \, y_1 \, y_0^0 + x_1^0 \, x_0 \, y_1 \, y_0 + x_1 \, x_0^0 \, y_1^0 \, y_0 +$$
$$x_1 \, x_0^0 \, y_1 \, y_0 + x_1 \, x_0 \, y_1^0 \, y_0 + x_1 \, x_0 \, y_1 \, y_0^0,$$
$$(MUL_{2,2})_1 = (x_1 + x_0 + y_1 + y_0)\,(x_1 + x_0 + y_1 + y_0^0)\,(x_1 + x_0 + y_1^0 + y_0)$$
$$(x_1 + x_0 + y_1^0 + y_0^0)\,(x_1 + x_0^0 + y_1 + y_0)\,(x_1 + x_0^0 + y_1 + y_0^0)$$
$$(x_1^0 + x_0 + y_1 + y_0)\,(x_1^0 + x_0 + y_1^0 + y_0)\,(x_1^0 + x_0^0 + y_1 + y_0)$$
$$(x_1^0 + x_0^0 + y_1^0 + y_0^0).$$

$\diamond$

As a particular consequence of this observation, each switching function $f \in \mathbb{B}_n$ can be described both in terms of a disjunctive normal form and in terms of a conjunctive normal form.

Example 4.9. The parity function $PAR_5 \in \mathbb{B}_5$,

$$PAR_5(x_1, \ldots, x_5) = x_1 \oplus \ldots \oplus x_5 \quad \left(= \sum_{i=1}^{5} x_i \pmod 2 \right),$$

has the following canonical disjunctive normal form:

$$
\begin{aligned}
PAR_5(x_1, \ldots, x_5) = \;\; & x_1 \, \overline{x_2} \, \overline{x_3} \, \overline{x_4} \, \overline{x_5} \;+\; \overline{x_1} \, x_2 \, \overline{x_3} \, \overline{x_4} \, \overline{x_5} \;+\; \overline{x_1} \, \overline{x_2} \, x_3 \, \overline{x_4} \, \overline{x_5} \\
& +\; \overline{x_1} \, \overline{x_2} \, \overline{x_3} \, x_4 \, \overline{x_5} \;+\; \overline{x_1} \, \overline{x_2} \, \overline{x_3} \, \overline{x_4} \, x_5 \\
& +\; x_1 \, x_2 \, x_3 \, \overline{x_4} \, \overline{x_5} \;+\; x_1 \, x_2 \, \overline{x_3} \, x_4 \, \overline{x_5} \;+\; x_1 \, x_2 \, \overline{x_3} \, \overline{x_4} \, x_5 \\
& +\; x_1 \, \overline{x_2} \, x_3 \, x_4 \, \overline{x_5} \;+\; x_1 \, \overline{x_2} \, x_3 \, \overline{x_4} \, x_5 \;+\; x_1 \, \overline{x_2} \, \overline{x_3} \, x_4 \, x_5 \\
& +\; \overline{x_1} \, x_2 \, x_3 \, x_4 \, \overline{x_5} \;+\; \overline{x_1} \, x_2 \, x_3 \, \overline{x_4} \, x_5 \;+\; \overline{x_1} \, x_2 \, \overline{x_3} \, x_4 \, x_5 \\
& +\; \overline{x_1} \, \overline{x_2} \, x_3 \, x_4 \, x_5 \\
& +\; x_1 \, x_2 \, x_3 \, x_4 \, x_5.
\end{aligned}
$$

This representation consists of $\binom{5}{1} + \binom{5}{3} + \binom{5}{5} = 16$ monomials. $\diamond$

Although conjunctive and disjunctive normal forms have many pleasant properties, at the moment there are insurmountable problems concerning their algorithmic handling. Here, the main source of difficulties can be seen in the fact that neither disjunctive nor conjunctive normal forms of a switching function are uniquely determined.

Definition 4.10. *A representation type of switching functions is called* **canonical** *if each switching function f has exactly one representation of this type. A representation type is called* **universal** *if for each switching function there exists a representation of this type.*

Example 4.11. Truth tables, canonical disjunctive normal forms, and canonical conjunctive normal forms constitute a canonical representation of switching functions. However, neither the set of all disjunctive normal forms nor the set of all conjunctive normal forms constitutes a canonical representation, as there may be many different such normal forms for a given function f.   ◇

The property of a representation type that it is canonical plays an essential role in its algorithmic behavior. Namely, if each function has exactly one representation, then it can be tested quite easily whether two given functions coincide: one has merely to test whether their representations are identical. The fact that neither disjunctive normal forms nor conjunctive normal forms provide a canonical representation is the main reason why equivalence checking is difficult in this case. In the following, we prove that efficient algorithms for solving equivalence problems of disjunctive or conjunctive normal forms cannot be expected, as both problems are **co-NP**-complete. Our proof is based on the classic result that 3-SAT is **NP**-complete.

Problem 4.12. The problem **3-SAT** is defined as follows:

Input: A conjunctive normal form in which each clause consists of exactly 3 literals.

Output: "Yes", if there exists a satisfying assignment for the represented function. "No", otherwise.

The following theorem is a central statement of the theory of **NP**-completeness that was introduced in Section 2.5.

Theorem 4.13. *The problem 3-SAT is* **NP***-complete.*   □

Incidentally, one part of the theorem, namely the claimed membership of 3-SAT in the class **NP** is trivial. One "guesses" a satisfying assignment, and then applies substitutions in order to verify that all clauses are satisfied for this input (which is obviously possible in polynomial time).

Problem 4.14. Let X be a representation type of switching functions. The problem **EQU$_X$** is defined as follows:

Input: Two representations P_1 and P_2 from the class given by X.

Output: "Yes", if P_1 and P_2 represent the same function. "No", otherwise.

Theorem 4.15. *The equivalence test EQU$_{DNF}$ of two disjunctive normal forms is co-NP-complete. The same statement holds for the equivalence test EQU$_{CNF}$ of two conjunctive normal forms.*

Proof. We have to show that the test whether two disjunctive or conjunctive normal forms represent *different* functions is **NP**-complete.

Membership in NP: First, one "guesses" an assignment which implies different function values for the two disjunctive (conjunctive) normal forms. By evaluating the functions for the chosen assignments, this difference can be verified in polynomial time.

3-SAT $\leq_P$ EQU$_{CNF}$: Let C be an input of 3-SAT, i.e., C is a conjunctive normal form all of whose clauses consists of exactly three literals. We choose $C_1 = C$ and $C_2 = 0$. (In particular, C_2 is the trivial conjunctive normal form of the contradiction function.) Now C can be satisfied if and only if C_1 and C_2 represent different functions.

3-SAT $\leq_P$ EQU$_{DNF}$: Here, let $C = \prod_{i=1}^{m} c_i$ be an input of 3-SAT. Each clause c_i of C is of the form

$$c_i = x_{i1}^{e_{i1}} + x_{i2}^{e_{i2}} + x_{i3}^{e_{i3}},$$

with $e_{ij} \in \{0,1\}$, $1 \leq i \leq m$, $1 \leq j \leq 3$. The function f_C defined by C satisfies

$$f_C = c_1 c_2 \ldots c_m.$$

Due to DeMorgan's rules, f_C can be satisfied if and only if

$$\overline{f_C} = \overline{c_1} + \overline{c_2} + \ldots + \overline{c_m}$$

is not the tautology function.

Further, let $k_i = \overline{c_i}$, $1 \leq i \leq m$. Another application of DeMorgan's rules yields

$$k_i = \overline{c_i} = x_{i1}^{1-e_{i1}} x_{i2}^{1-e_{i2}} x_{i3}^{1-e_{i3}}, \qquad 1 \leq i \leq m.$$

Let D_1 be the disjunctive normal form with monomials $k_1, \ldots, k_m$, and let D_2 be the disjunctive normal form of the tautology function, $D_2 = 1$. Then C can be satisfied if and only if D_1 and D_2 represent different functions. $\square$

Sometimes, the very modest algebraic structure within the two operations $+$ and $\cdot$ makes it hard to work with disjunctive and conjunctive normal forms. For example, neither the equation $f + g = h$ nor the equation $f \cdot g = h$ can be solved for f. By employing the operations $(\oplus, \cdot)$ instead of the pair $(+, \cdot)$, the situation changes fundamentally: the set $\mathbb{B} = \{0, 1\}$ in connection with the operations $\oplus$ and $\cdot$ constitutes a field in the algebraic sense. This fact allows us to apply the extensive tool set from the theory of algebraic fields in the context of switching functions. Parity normal forms are also of particular interest for practical applications. As the $\oplus$-operation will change its function value whenever exactly one bit is modified, this operation serves as a good basis for designing circuits that can be tested easily.

Definition 4.16. *A* **ring sum expansion** *(RSE) or (positive polarity)* **Reed-Muller expansion** *is a parity normal form where all monomials only include positive literals.*

The next theorem states an important structural property for ring sum expansions.

Theorem 4.17. *For each switching function $f \in \mathbb{B}_n$, there exists a uniquely determined representation in terms of a ring sum expansion. Hence, ring sum expansions are a canonical representation.*

Proof. Let $f \in \mathbb{B}_n$. First we prove the *existence* of a corresponding ring sum expansion. We start from the canonical DNF representation of f,

$$f(x_1, \ldots, x_n) = \sum_{c \in \mathrm{on}(f)} m_c(x_1, \ldots, x_n).$$

For each vector $c \in \mathrm{on}(f)$ exactly one minterm is 1 in this DNF. As 1 is an odd number, we have

$$f(x_1, \ldots, x_n) = \bigoplus_{c \in \mathrm{on}(f)} m_c(x_1, \ldots, x_n).$$

This a parity normal form of f. Now we have to remove negative literals from this normal form. This is possible by using the equation $\overline{y} = y \oplus 1$: each negative literal $\overline{x_i}$ is replaced by the expression $(x_i \oplus 1)$. Then the resulting expression is expanded by using distributivity, associativity, and the rules for computations over an algebraic field. Finally, we obtain the ring sum expansion

$$f(x_1, \ldots, x_n) = \bigoplus_{I \in \mathcal{I}} m_I,$$

where $m_I = x_{i_1} \ldots x_{i_s}$ if $I = \{i_1, \ldots, i_s\} \in \mathcal{I}$, and $\mathcal{I}$ is a suitable subset $\mathcal{I} \subset 2^{\{1, \ldots, n\}}$.

Now we prove *uniqueness*. For any $I \subset \{1, \ldots, n\}$, there is exactly one m_I. Hence, each ring sum expansion is characterized uniquely by a set $\mathcal{I} \subset 2^{\{1, \ldots, n\}}$. As the number of subsets of $2^{\{1, \ldots, n\}}$ is exactly 2^{2^n}, the number of possible ring sum expansions coincides with the number of switching functions in n variables. As each switching function can be represented in terms of a ring sum expansion, this one must be unique. □

Example 4.18. In this example, we illustrate the conversion from the previous proof by constructing a ring sum expansion from the disjunctive normal form $f(x_1, x_2, x_3) = x_1 \overline{x_2} x_3 + \overline{x_1} x_2 \overline{x_3}$.

$$
\begin{aligned}
f(x_1, x_2, x_3) &= x_1 \overline{x_2} x_3 + \overline{x_1} x_2 \overline{x_3} \\
&= x_1 \overline{x_2} x_3 \oplus \overline{x_1} x_2 \overline{x_3} \\
&= x_1 (x_2 \oplus 1) x_3 \oplus (x_1 \oplus 1) x_2 (x_3 \oplus 1) \\
&= x_1 x_2 x_3 \oplus x_1 x_3 \oplus x_1 x_2 x_3 \oplus x_2 x_3 \oplus x_1 x_2 \oplus x_2 \\
&= x_1 x_3 \oplus x_2 x_3 \oplus x_1 x_2 \oplus x_2.
\end{aligned}
$$

◇

Due to uniqueness of the representation, it can be checked easily if two given ring sum expansions describe the same function. A severe drawback of ring sum expansions is the fact that even some simple functions like the disjunction of n variables require representations of exponential length.

Theorem 4.19. *For a set of indices* $I = \{i_1, \ldots, i_s\}$ *let* $m_I = x_{i_1} \ldots x_{i_s}$. *The ring sum expansion of*

$$
f(x_1, \ldots, x_n) = x_1 + x_2 + \ldots + x_n
$$

is

$$
f(x_1, \ldots, x_n) = \bigoplus_{I \subset \{1, \ldots, n\} \setminus \emptyset} m_I, \tag{4.1}
$$

and hence of exponential length.

Proof. We prove that the given ring sum expansion (4.1) computes the disjunction of n variables. The claim then follows from the uniqueness of ring sum expansions.

In case $x = (x_1, \ldots, x_n) = 0$, none of the $2^n - 1$ monomials (4.1) computes the value 1. Therefore, $f(x) = 0$.

Now let x be an input which consists of exactly l inputs for an $l > 0$. Then, for each $1 \le k \le l$, there are exactly

$$
\binom{l}{k}
$$

monomials in k variables that compute a 1. Consequently, the number of monomials computing a 1 is exactly

$$\sum_{k=0}^{l} \binom{l}{k} - \binom{l}{0} = 2^l - 1.$$

As for all $l > 0$ this value is odd, the parity function computes a 1. Hence, we have shown that the ring sum expansion (4.1) computes the disjunction of n variables.   □

4.3 Circuits and Formulas

4.3.1 Circuits

Turning from the extensively investigated and well understood two-level normal forms towards multiple-level forms, then in many cases substantially more compact representations of switching functions can be found. Multiple-level representations are based on iterated application of a fixed set of basis operations. More precisely, let I be an arbitrary (finite or infinite) index set and $\Omega = \{\omega_i \in \mathbb{B}_{n_i} : i \in I\}$ be a set of basic functions. Ω is called a **basis**. An Ω-**circuit** S in n variables denotes an acyclic graph with two types of nodes:

- **input nodes**, which are nodes without an incoming edge, labeled by 0, 1, or by a variable x_i.
- **function nodes**, which are nodes with indegree n_i, labeled by a basis function $\omega_i \in \Omega, i \in I$.

Each node v of the circuit S represents a switching function $f_{S,v}$ according to the following inductive rules:

- If v is labeled by 0 (1), then $f_{S,v}(x_1, \ldots, x_n) = 0$ (1).
- If v is labeled by x_i, then $f_{S,v}(x_1, \ldots, x_n) = x_i$.
- If v is labeled by ω_i and has predecessors $v_1, \ldots, v_{n_i}$, then $f_{S,v}(x_1, \ldots, x_n) = \omega_i(f_{S,v_1}(x), \ldots, f_{S,v_{n_i}}(x))$.

In general, we are only interested in some of the functions being represented within the graph. We mark the relevant nodes $v_1, \ldots, v_m$ as **output nodes**, and define the function $f_S \in \mathbb{B}_{n,m}$ which is represented by the circuit S by

$$f_S(x) = (f_{S,v_1}(x), \ldots, f_{S,v_m}(x)).$$

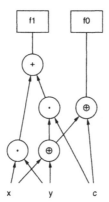

Figure 4.2. Circuit representation of a full adder

Example 4.20. A **full adder** $f = f(x, y, c) \in \mathbb{B}_{3,2}$, well-known as a basic hardware component in chip design, computes the binary sum $x + y + c$ of the three input bits x, y, and c. In a typical application, c is the carry bit of the previous addition. Let $\Omega = \{+, \cdot, \bar{}\}$ be called the **standard basis**. Figure 4.2 shows an Ω-circuit of f with output nodes f_1, f_0.   ◇

The number of operation levels in the circuit is called the **depth** of the circuit. In a simplified consideration of physical chips, the depth corresponds to the time consumption of the realized functionality.

Definition 4.21. *A basis Ω is called* **complete** *if Ω-circuits provide a universal representation type, i.e., if each switching function $f \in \bigcup_{n \geq 0} \mathbb{B}_n$ can be represented in terms of an Ω-circuit.*

Example 4.22. (a) The standard basis $\{+, \cdot, \bar{}\}$ is complete, as each switching function can be represented in terms of a disjunctive normal form, which can be seen as a circuit of depth 3 over the standard basis. The three operations are often symbolized by the nodes in Fig. 4.3, which are usually called **gates**.

(b) The basis $\{\oplus, \cdot\}$ is complete, as each switching function can be represented in terms of a ring sum expansion.

(c) The two bases $\{\cdot, \bar{}\}$ and $\{+, \bar{}\}$ are complete, as disjunction (resp. conjunction) can be expressed in terms of conjunction (resp. disjunction) by means of DeMorgan's rule and negation.

(d) The set $\{+, \cdot\}$ is not a complete basis, as it can only generate monotone increasing functions.   ◇

$$f = f_1 + f_2 \qquad\qquad f = f_1 \cdot f_2 \qquad\qquad f = \overline{f_1}$$

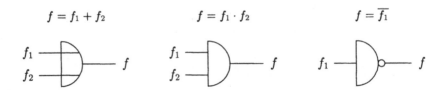

Figure 4.3. Symbolic representations of disjunction, conjunction, and complementation

One immediately recognizes that even for a fixed basis, circuit representations of switching functions are not uniquely determined, i.e., circuits are not a canonical representation type. As in the case of two-level normal forms, we are particularly interested in representations which consist of few nodes, as this corresponds to a chip realization with few gates. The (Ω-)**circuit complexity** of a switching function denotes the number of nodes in a minimal (Ω-)circuit representation of this function.

As disjunctive and conjunctive normal forms are specific circuits, the circuit complexity of a switching function over the standard basis is never greater than its minimal DNF or CNF representation. On the contrary, in most cases circuit representations are substantially more compact.

This advantage of the circuit model is contrasted by a severe drawback, like in the case of disjunctive or conjunctive normal forms: the enormous time complexity of checking circuits for equivalence.

Theorem 4.23. *The equivalence test of Boolean circuits over the standard basis* $\{+, \cdot, ^-\}$ *is* **co-NP***-complete.*

Proof. As the function value of a circuit can be evaluated in polynomial time, the problem is in **co-NP**.

From Theorem 4.15 it is known that the problem EQU$_{\mathrm{DNF}}$ is **co-NP**-complete. This problem can be immediately reduced to the equivalence problem of Boolean circuits, as each conjunction and each disjunction can be realized by a corresponding circuit node. $\qquad\square$

4.3.2 Formulas

A typical property of circuits is their ability to use a function $f_{S,v}$ represented in a node v as input to other nodes more than once. In other words, in the case of circuits the outdegree of v can be greater than 1. Often it is exactly this property that allows very compact representations. However, as this property makes algorithmic handling significantly harder, the restricted model of so-called *formulas* has been studied intensively. Here, multiple use of intermediate functions within the graph is not allowed.

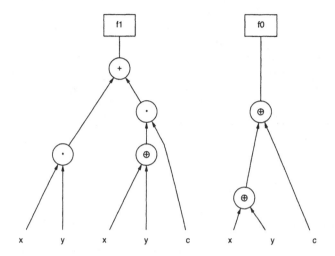

Figure 4.4. Formula representation of a full adder

Definition 4.24. *Let Ω be a basis. An Ω-formula is a circuit whose nodes have outdegree 1.*

Consequently, an Ω-circuit is a formula if and only if the graph consists of individual trees. A formula representation of the full adder from Example 4.20 is depicted in Fig. 4.4.

Due to space limitations we can only give a reference to the interesting bijective connection between Ω-formulas over the standard basis $\Omega = \{+, \cdot, ^-\}$ and Boolean formulas of the switching algebra from Section 3.2. Surely, it is also due to this connection that so much research effort has been devoted to the investigation of formulas.

Although Ω-formulas make the analysis of the represented functions easier – for example, when proving lower bounds for the representation size of concrete functions – the equivalence test of two Boolean formulas is **co-NP**-complete like in the case of circuits.

4.4 Binary Decision Trees and Diagrams

In the previous section, we introduced several representations describing switching functions by means of a *computation process*. In contrast, this section deals with representation types which describe a switching function by means of an *evaluation process*.

In natural sciences, evaluation processes serve for the classification of objects. Depending on the results of several tests, an object is put into a specific class.

A good illustration of such evaluation processes can be obtained by means of directed graphs, so-called **decision diagrams**. Here, each node is associated with exactly one classifying test. The consecutive order of the single tests and the evaluation of the tests is described by means of edges. For switching functions, the tests may be of the form: *Is the i-th input bit x_i zero, or one ?* In case of the answer "$x_i = 0$", one can proceed with different tests than in case "$x_i = 1$".

In the following, we are concerned with representations which are based on such decision processes. It will turn out that those representation types are particularly useful in the context of Boolean manipulation.

4.4.1 Binary Decision Trees

A **binary decision tree** in n variables is a tree, whose

- internal nodes are labeled by a variable x_i and have exactly two outgoing edges, and whose
- sinks are labeled by one of the two constants 0 or 1.

A decision tree represents a switching function $f \in \mathbb{B}_n$ in a natural way. Each assignment to the variables $x_1, \ldots, x_n$ defines a uniquely determined path from the root of the tree to a sink. The label of this sink gives the value of the function on this input.

We can always assume that each variable is read at most once on each path of the decision tree. A second evaluation of an already tested variable does not partition the input set any further: if all inputs have left the first test along the 0-edge (1-edge), then they also leave the next test along the 0-edge (1-edge). The other edge together with the subtree rooted in the successor node are never visited and can therefore be eliminated.

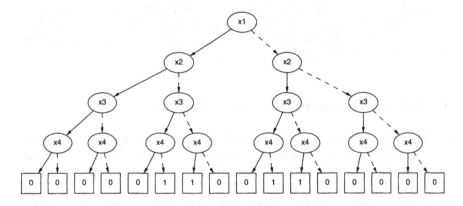

Figure 4.5. Complete decision tree of the function $f = (x_1 \oplus x_2) \cdot (x_3 \oplus x_4)$

Example 4.25. By iterated application of Shannon's expansion from Theorem 3.21 and "graphical recording" we obtain a complete decision tree of $f = (x_1 \oplus x_2) \cdot (x_3 \oplus x_4)$, see Fig. 4.5. In the diagrams, 1-edges are drawn as solid arrows, and 0-edges will be drawn as dashed arrows.

In many cases, there exist significantly more compact decision tree representations than those by means of the complete decision tree. An example for the function f is shown in Fig. 4.6.   ◇

Often, by changing the order of the partitioning variables in the individual subgraphs, new possibilities for minimizing the size of the decision tree can be revealed. This applies, e.g., to the tree in Fig. 4.7 which can also be transformed to the decision tree in Fig. 4.6. In general however, with regard

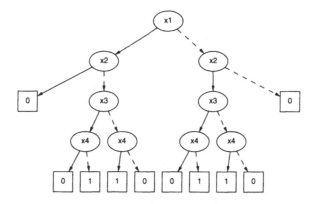

Figure 4.6. More compact decision tree of the function $f = (x_1 \oplus x_2) \cdot (x_3 \oplus x_4)$

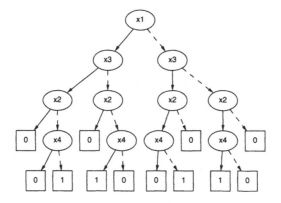

Figure 4.7. A decision tree of $f = (x_1 \oplus x_2) \cdot (x_3 \oplus x_4)$, where the variables are tested in the order x_1, x_3, x_2, x_4

to the representation size, decision trees are inferior to the model of branching programs discussed next.

4.4.2 Branching Programs

Branching programs can be interpreted as a generalization of binary decision trees where the tree structures are allowed to be replaced by general graph structures. Branching programs were first investigated by C. Lee in 1959, under the name *binary decision programs*. The original motivation was to find an alternative representation to Boolean circuits. A first systematic study of branching programs was presented by W. Masek in 1976.

A **branching program** or a **binary decision diagram** in n variables is a directed acyclic graph with exactly one root, whose

- sinks are labeled by the Boolean constants 0, 1, and whose
- internal nodes are labeled by a variable x_i and have exactly two outgoing edges, a 0-edge and a 1-edge.

Analogous to decision trees, branching programs represent a switching function $f \in \mathbb{B}_n$ in the following way. Each assignment to the input variables x_i defines a uniquely determined path from the root of the graph to one of the sinks. The label of the reached sink gives the function value on this input. The **size** of a branching program is the number of its nodes. Figure 4.8 shows two different branching programs of the parity function in four variables.

For decision trees we could assume – without restricting the model – that a variable x_i is tested on each path at most once. A second test of the variable

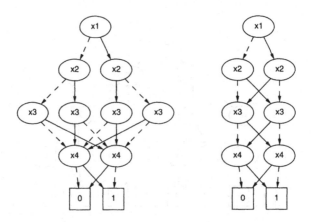

Figure 4.8. Two different branching programs of the parity function $x_1 \oplus x_2 \oplus x_3 \oplus x_4$

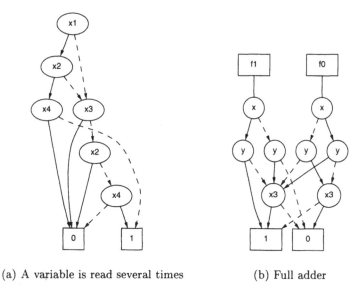

(a) A variable is read several times   (b) Full adder

Figure 4.9. Specific branching programs

on a path could be removed immediately, as this test had to be redundant. For branching programs this property does not hold. A node v in which a variable x_i is read can have more than one predecessor. For this reason, there may be several paths leading from the root to the node v. On these different paths from the root to v, the variables may be tested in a different manner. Hence, even if the variable x_i has already been tested on a path from the root to v, the test in v cannot be eliminated in general.

Example 4.26. In the branching program in Fig. 4.9 (a), there exists a path from the root to a sink on which the variable x_2 is tested more than once. Nevertheless, the node containing the second test is *not* redundant for the whole branching program.   ◇

Branching programs can also be used for compactly representing Boolean functions $f \in \mathbb{B}_{n,m}$ with several outputs. Besides the trivial possibility of representing each component f_i of f by its own branching program P_i, it is possible to combine all components of f within a single branching program with m distinguished output nodes. Figure 4.9 (b) demonstrates this possibility in the case of the full adder from Example 4.20.

Now we would like to investigate the problem of how branching programs and other representation types can be transformed to or from each other. However, we do not discuss the exact details of a possible implementation, which may involve suitable adjacency lists. The algorithm in Fig. 4.10 serves for

Input: A branching program P of $f \in \mathbb{B}_n$.
Output: DNF of f.

Algorithm: Traverse all paths in P that start in the root and lead to the 1-sink. With each of these paths, associate a monomial as follows. Let $x_{i_1}, \ldots, x_{i_k}$ be the labels of the nodes on the path, and let $e_1, \ldots, e_k \in \{0, 1\}$ be the corresponding values of the traversed edges. Then the associated monomial is

$$x_{i_1}^{e_1} \cdot \ldots \cdot x_{i_k}^{e_k}.$$

The disjunction of these monomials over all paths yields a DNF of f.

Figure 4.10. Transformation of a branching program into a DNF

computing a disjunctive normal form from a given branching program. Here, for each path from the root to the 1-sink the corresponding minterm is determined. Note that for multiple occurrences of a variable, the corresponding monomial may collapse to the zero function.

Branching programs can be transformed quite easily into circuits with respect to a complete basis Ω while preserving the structure of the branching program. The transformation algorithm is shown in Fig. 4.11, and an example of the transformation is given in Fig. 4.12.

At the end of this section we discuss some properties of branching programs.

Definition 4.27. *Let P be a branching program.*

1. *The k-th* **level** *of P denotes the set of all nodes which can be reached from the root by a path of length $k - 1$.*

2. *The maximal cardinality $w(P)$ of a level of P is called the* **width** *of P.*

Input: A branching program P of $f \in \mathbb{B}_n$.
Output: An Ω-circuit of f with respect to a complete basis Ω.

Algorithm: Let $sel \in \mathbb{B}_3$ be the switching function

$$sel(x, y, z) = \overline{x} \cdot y + x \cdot z$$

which represents a selection process. First, we transform P into a $\{sel\}$-circuit P'. To each internal node v in the graph P, we add a new predecessor node which is labeled by the variable of v. Replace the label of v by the function sel. Change the directions of all edges and forget their labels.

As Ω is a complete basis, there exists an Ω-circuit C_{sel} which realizes the sel-function. Realize all occurrences of sel in P' by the Ω-circuit C_{sel}.

Figure 4.11. Transformation of a branching program into an Ω-circuit

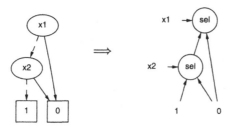

Figure 4.12. Transformation of a branching program into a *sel*-circuit

By using these notions, some important special cases of branching programs can be defined.

Definition 4.28. *Let P be a branching program.*

1. *P is called* **width-bounded** *with width k if each level is of cardinality at most k.*

2. *P is called* **synchronous** *if for each node v of P, all paths from the root to v are of equal length.*

3. *P is called* **oblivious** *if P is synchronous and all non-sink nodes within a level are labeled by the same variable.*

4. *P is a* **read-k-times-only branching program** *if each variable occurs on every path at most k times.*

These definitions imply the following subclasses of branching programs:

$$\mathbf{BP_{width\text{-}k}} = \{ \text{ branching programs of width } \leq k \ \},$$
$$\mathbf{sBP} = \{ \text{ synchronous branching programs } \},$$
$$\mathbf{oBP} = \{ \text{ oblivious branching programs } \},$$
$$\mathbf{BP}k = \{ \text{ read-k-times-only branching programs } \}.$$

4.4.3 Read-Once Branching Programs

In the previous two sections we have discussed decision trees and branching programs. There are two essential differences between these two representation types:

- In contrast to decision trees, branching programs are allowed to read a variable more than once on a path.

- In contrast to the nodes in decision trees, the nodes in branching programs may have more than one predecessor.

With regard to the algorithmic properties, we will show that the efficiency of important algorithms can only be guaranteed if none of the variables is read several times on a path. Branching programs with this property are called **read-once branching programs**. The model of read-once branching programs was already studied by A. Cobham in 1966, and later by W. Masek in 1976. This model is also of great interest from the theoretical point of view, as since then a lot of powerful techniques for proving lower bounds concerning the representation size of concrete functions have been developed. Incidentally, the definitions of read-once branching programs and decision trees imply that each decision tree is also a read-once branching program.

An important structural property of read-once branching programs is as follows. Each path in a read-once branching program is a **computation path**, i.e., for all paths from the root to a sink there exists an input vector with the property that the edge values within the graph (which are the answers to the tests on the path) coincide with the corresponding components of the input vector. For arbitrary branching programs this property does not hold, as a path may contain both the 0-edge of a node with label x_i and the 1-edge of another node with label x_i.

The branching programs in Figs. 4.5, 4.6, 4.8, and 4.9 (b) are read-once branching programs, but the branching program in Fig. 4.9 (a) does not have this property.

4.4.4 Complexity of Basic Operations

Now we analyze the complexity of some basic operations for different types of branching programs. This shows that each of the individual types provably suffers from difficulties in algorithmic handling. Interestingly enough, these difficulties occur in different situations.

Problem 4.29. The problem **3-SAT/3-OCCURRENCES** is defined as follows:

Input: A conjunctive normal form C whose clauses consist of exactly three literals, and which has the property that each variable occurs at most three times in C.

Output: "Yes", if there is a satisfying assignment for the represented function. "No", otherwise.

Lemma 4.30. *The problem 3-SAT/3-OCCURRENCES is* **NP***-complete.*

Proof. As a given assignment can be verified easily for satisfiability, 3-SAT/3-OCCURRENCES is in **NP**.

Now we give a polynomial time reduction from 3-SAT to the problem 3-SAT/3-OCCURRENCES. Let $C = \prod_{i=1}^{m} c_i$ and $c_i = x_{i1}^{e_{i1}} + x_{i2}^{e_{i2}} + x_{i3}^{e_{i3}}$, where $e_{ij} \in \{0,1\}$, $1 \leq i \leq m$, $1 \leq j \leq 3$.

We construct a new conjunctive normal form C' by replacing each variable x_i in C by a new variable, $1 \leq i \leq n$. The new variable for the j-th occurrence of x_i is denoted by x_{ij}. Additionally, for all $1 \leq i \leq n$ the following clauses are added:

$$(x_{i1} + \overline{x_{i2}}) \cdot (x_{i2} + \overline{x_{i3}}) \cdot \ldots \cdot (x_{i,k_i-1} + \overline{x_{i,k_i}}) \cdot (x_{i,k_i} + \overline{x_{i1}}), \qquad (4.2)$$

where k_i denotes the number of occurrences of variables x_i.

Thus, for example, the 3-SAT formula

$$C = (x_1 + \overline{x_2} + x_4) \cdot (x_2 + x_3 + \overline{x_4})$$

is transformed into the formula

$$C' = (x_{11} + \overline{x_{21}} + x_{41}) \cdot (x_{22} + x_{31} + \overline{x_{42}})$$
$$\cdot (x_{11} + \overline{x_{11}})$$
$$\cdot (x_{21} + \overline{x_{22}}) \cdot (\overline{x_{21}} + x_{22})$$
$$\cdot (x_{31} + \overline{x_{31}}).$$

Obviously, if $(x_1, \ldots, x_n)$ is a satisfying assignment of C, then the assignment

$$x_{ij} = x_i, \qquad 1 \leq j \leq k_i, \qquad 1 \leq i \leq n$$

is a satisfying assignment of C'.

Now let us consider a satisfying assignment of C'. If this one includes $x_{i1} = 0$, then relation (4.2) implies successively $x_{i2} = 0$, $x_{i3} = 0$, ..., $x_{i,k_i} = 0$. If $x_{i1} = 1$, then relation (4.2) implies successively $x_{i,k_i} = 1$, $x_{i,k_i-1} = 1$, ..., $x_{i,2} = 1$. As a consequence, we always have $x_{i1} = x_{i2} = \ldots = x_{i,k_i}$. This observation immediately implies that the assignment

$$x_i = x_{i1}, \qquad 1 \leq i \leq n$$

satisfies the formula C. Hence, each algorithm for solving 3-SAT/3-OCCURRENCES can also be used for solving 3-SAT.

Of course, the described transformation can be performed in polynomial time.

□

Problem 4.31. Let X be a representation type of switching functions. The problem **SAT$_X$** is defined as follows:

Input: A representation P of type X.

Output: "Yes", if there exists a satisfying assignment of the function which is represented by P. "No", otherwise.

We show that the satisfiability test is already hard for quite specific subclasses of branching programs.

Theorem 4.32. *The problems* $SAT_{width\text{-}2\ sBP3}$ *and* $SAT_{width\text{-}3\ oBP2}$ *are* both **NP**-*complete.*

Proof. To prove the first statement we reduce the **NP**-complete problem 3-SAT/3-OCCURRENCES to the problem $SAT_{width\text{-}2\ sBP3}$. Let $C = \prod_{i=1}^{m} c_i$ and $c_i = x_{i1}^{e_{i1}} + x_{i2}^{e_{i2}} + x_{i3}^{e_{i3}}$, where $e_{ij} \in \{0, 1\}$, $1 \leq i \leq m$, $1 \leq j \leq 3$. The literals $x_i^1 = x_i$ and $x_i^0 = \overline{x_i}$ can be represented in terms of simple branching programs which exactly consist of a root node and two sinks. The conjunction of two branching programs P_1 and P_2 can be performed by identifying the 1-sinks of P_1 with the root of P_2. In this way, branching programs of the clauses c_i can easily be constructed. The disjunction of P_1 and P_2 can be performed by identifying the 0-sinks of P_1 with the root of P_2. Using this construction, a branching program $P(C)$ of the conjunctive normal form C can be generated (see Fig. 4.13 (a)). It is not difficult to construct $P(C)$ as a synchronous branching program of width 2. Figure 4.13 (b) illustrates this

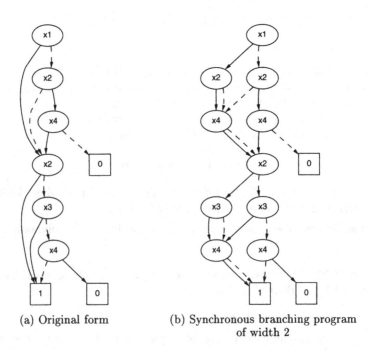

(a) Original form   (b) Synchronous branching program
                                of width 2

Figure 4.13. Constructing branching programs of the 3-SAT formula $(x_1 + \overline{x_2} + x_4) \cdot (x_2 + x_3 + \overline{x_4})$

fact in case of the function $(x_1 + \overline{x_2} + x_4) \cdot (x_2 + x_3 + \overline{x_4})$. As each variable occurs at most three times in the input of 3-SAT/3-OCCURRENCES, the branching program $P(C)$ is in the class BP3.

The construction yields a polynomial time reduction from 3-SAT/3-OCCUR-RENCES to SAT$_{\text{width-2 sBP3}}$. It remains to show that SAT$_{\text{width-2 sBP3}}$ is in **NP**. However, this is immediately clear, as the result of a branching program for a particular input can be evaluated in polynomial time.

To prove the second statement, we modify the reduction used for proving the **NP**-completeness of SAT$_{\text{width-2 sBP3}}$. First, $P(C)$ can be easily transformed into an oblivious branching program of width 3. This only requires adding some nodes whose 1- and 0-successors are identical. The construction is illustrated in Fig. 4.14 (a).

Then each occurrence of a variable in $P(C)$ is replaced by a new variable, which results in a new oblivious read-once branching program of width 3 denoted by $P'(C)$. Furthermore, we can easily build an oblivious read-once branching program of width 3 that tests the equality of the variables $x_1, \ldots, x_k$ (see Fig. 4.14 (b)). Hence, we can construct a width-3 oBP1 called

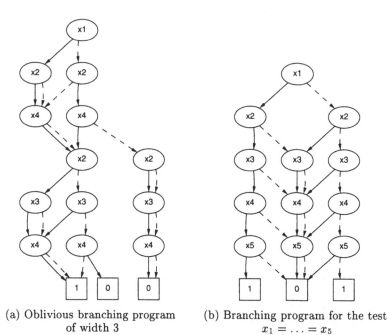

(a) Oblivious branching program of width 3

(b) Branching program for the test $x_1 = \ldots = x_5$

Figure 4.14. Construction of branching programs for proving **NP**-completeness of SAT$_{\text{width-3 oBP2}}$

$P''(C)$, which checks whether $x = y$ for all x, y in $P'(C)$ that correspond to the same variable of $P(C)$.

$P(C)$ can be satisfied if and only if there is a variable assignment for which $P'(C)$ and $P''(C)$ simultaneously compute the value 1. As the conjunction of $P'(C)$ and $P''(C)$ leads to an oblivious read-2 branching program of width 3, the desired polynomial time reduction from 3-SAT/3-OCCURRENCES to $SAT_{\text{width-3 oBP3}}$ has been established.   □

In particular, we can easily deduce the following corollary:

Corollary 4.33.   *1. The problems SAT_{BP2}, SAT_{oBP}, SAT_{sBP}, as well as SAT_{BP} are **NP**-complete.*

  *2. The problems EQU_{BP2}, EQU_{oBP}, EQU_{sBP}, as well as EQU_{BP} are **co-NP**-complete.*

Proof. The second statement follows immediately from the observation

$$\neg \, SAT_X(P) = EQU_X(P, 0),$$

where P is a representation of type X for an arbitrary switching function, and $\neg P$ denotes the complement of a problem $\mathcal{P}$.   □

Problem 4.34. The problem **COMMON_PATH$_{BP1}$** is defined as follows:

Input: Two read-once branching programs P_1 and P_2.

Output: "Yes", if there exists an assignment to the input variables such that P_1 and P_2 simultaneously compute the value 1. "No", otherwise.

Lemma 4.35. *The problem COMMON_PATH$_{BP1}$ is **NP**-complete.*

Proof. As each of the two branching programs can be evaluated in polynomial time, the problem is in **NP**.

In the proof of Theorem 4.32, for each instance C of the problem 3-SAT/3-OCCURRENCES we have constructed two read-once branching programs $P'(C)$ and $P''(C)$ with the following property: C can be satisfied if and only if there exists a variable assignment for $P'(C)$ and $P''(C)$ such that $P'(C)$ and $P''(C)$ simultaneously compute the value 1. For this reason, the **NP**-completeness of 3-SAT/3-OCCURRENCES immediately implies the **NP**-completeness of COMMON_PATH$_{BP1}$.   □

Lemma 4.36. *The problem SAT_{BP1} can be solved in polynomial time.*

Proof. The property that each path in a read-once branching program leading from the root to a sink is a computation path immediately provides a procedure for deciding the satisfiability problem. By using depth first search, all nodes are marked that can be reached by a path starting in the root. P can be satisfied if and only if the 1-sink is marked in this step. $\square$

Problem 4.37. Let X be a representation type of switching functions, and let $*$ be a Boolean operation. The problem **$*$-SYN$_X$** is defined as follows:

Input: Two representations P_1 and P_2 of switching functions which are of type X.

Output: A representation of type X describing the function $f_1 * f_2$, where f_1 and f_2 are the functions being represented by P_1 and P_2.

In the proof of Theorem 4.32 we have seen that the problems $*$-SYN$_{oBP}$, $*$-SYN$_{sBP}$, and $*$-SYN$_{BP}$ for $* \in \{\cdot, +\}$ can be solved very easily. The same can be shown for all other Boolean operations. Interestingly enough, difficulties arise in case of read-once branching programs which are well behaved for other algorithmic problems.

Theorem 4.38. *The problem $\cdot$-SYN_{BP1} is **NP**-hard.*

Proof. The problem COMMON_PATH$_{BP1}$ can be solved by applying an algorithm for $\cdot$-SYN$_{BP1}$ to the input programs, and then deciding SAT$_{BP1}$ on the resulting program. According to Lemma 4.35 the problem COMMON_PATH$_{BP1}$ is **NP**-complete, and according to Lemma 4.36 the problem SAT$_{BP1}$ is solvable in polynomial time. Consequently, $\cdot$-SYN$_{BP1}$ is an **NP**-hard problem. $\square$

As complementation of read-once branching programs merely requires exchanging the two sink values, Theorem 4.38 can be immediately generalized to other Boolean operations.

Corollary 4.39. *The problems $+$-SYN_{BP1} and $\oplus$-SYN_{BP1} are **NP**-hard.*
$\square$

4.5 References

Concerning the historic works of Lee, Masek, and Cobham, the references are [Lee59, Mas76, Cob66].

More material on the mentioned classical representations of switching functions can be found in, e.g., [Weg87, Mei89]. Reference [Mor92] contains a survey on decision trees.

The complexity results for the different classes of branching programs go back to Gergov and Meinel [GM94b].

5. Requirements on Data Structures in Formal Circuit Verification

In medias res [Into the medium of things].
Horace (65–8 BC)

In the previous chapter, we have introduced several representation types for switching functions. In particular, we have described the switching functions under consideration based on the following ideas:

- by systematically tabulating the function values (e.g., truth tables),
- by providing a method for computing the function (e.g., disjunctive or conjunctive normal forms, Boolean formulas or circuits),
- by describing a scheme for evaluating the function (e.g., decision trees, branching programs).

Already at first glance, many differences between these representation types are apparent. We only mention some of them:

- The representation types differ substantially in size.

   Example 5.1. A truth table of a function $f \in \mathbb{B}_n$ consists always of 2^n rows. The size of all other treated representations depends on the particular function f. For the disjunction of the variables $x_1, \ldots, x_n$, for example, there exists a very compact disjunctive normal form

$$f(x) = x_1 + \ldots + x_n.$$

   On the other hand, according to Theorem 4.19, the ring sum expansion of this disjunction consists of $2^n - 1$ monomials. $\diamond$

- The representation types differ substantially in the efforts needed to evaluate the function value for a particular input.

   Example 5.2. In a truth table the function value can be determined by an easy look-up, whereas in the model of Boolean formulas the whole formula has to be evaluated for the particular input. $\diamond$

- The representation types differ substantially in the amount of computations necessary for representing the result of performing a Boolean operation on two functions.

Example 5.3. Let $f, g \in \mathbb{B}_n$, and let $*$ be an arbitrary Boolean operation. In case of circuits over a complete basis, the circuit representation of $f * g$ can be generated easily. The outputs of the circuits of f and g are used as inputs to a circuit that computes the operation $*$ (which exists because the basis is complete). In contrast to this simple construction, Theorem 4.38 shows that computing Boolean operations for read-once branching programs is **NP**-hard.   ◇

- The representation types differ substantially with regard to the complexity of specific properties.

Example 5.4. We consider the test whether a represented function f is constant. For read-once branching programs this test can be carried out easily: f is constant if and only if merely one of the two sinks can be reached from the root. On the other hand, the problem is **co-NP**-complete for conjunctive normal forms. The statement can be proven by using a simple reduction from the satisfiability problem of conjunctive normal forms, which is **NP**-complete according to Theorem 4.13: f cannot be satisfied if and only if both f is constant and $f(0, \ldots, 0) = 0$.   ◇

- The representation types differ substantially in the efforts needed for checking whether a variable is essential for a function.

Example 5.5. On the one hand, the uniqueness of ring sum expansions implies: a variable is essential for a function f if and only if it occurs in the ring sum expansion of f. On the other hand, the test of essentiality is **NP**-complete for conjunctive normal forms, as we have: a function f can be satisfied if and only if for a new variable x_0 the function $x_0 f$ depends essentially on the variable x_0.   ◇

The above examples show that each of the discussed representations simultaneously has advantages and drawbacks. For example, tabular representations allow a fast function evaluation for a given input. Due to their enormous size, the computation of a Boolean operation using truth tables is quite expensive. Contrariwise, in the case of circuits, Boolean operations can be performed fast, but the equivalence test is inherently difficult.

Although it is often possible to rank the different representation schemes with respect to different isolated aspects, it seems impossible to declare one of them to be the best with respect to all criteria. The essential reason for this is that in concrete applications the different criteria may be of different importance. In one application, memory usage may be the deciding factor, and in another application, it may be the running time of specific algorithms. For this reason, it is hopeless to expect a general measure for the best compromise without taking into account the specific application area.

In this book, we consider primarily applications in the field of computer-aided circuit design. In order to illustrate the use of switching functions in this area, we give a more detailed description of two paradigmatic applications from the hot topic of formal circuit verification. These applications have turned out to be particularly important, as they make the concrete requirements and demands on representations of switching functions quite precise. The problems to be solved occur not only in an isolated way, but often as subproblems in larger applications, too.

5.1 Circuit Verification

Before discussing the problem of circuit verification, let us review the objects to be verified, namely *digital circuits*. The first circuits in the historical development were employed to control electromechanical relays in telecommunication networks. These circuits were strictly **combinational** in the sense that they did not include memory elements. Of course, combinational circuits alone are only of limited interest. Their mode of operation can be improved substantially by adding memory elements which preserve the current state of the circuit. Circuits with memory elements are called **sequential circuits**. In spite of the greater functionality of sequential circuits, one must not underestimate the significance of combinational modules. Firstly, these modules compute functions "in one step," and secondly, each sequential circuit can be interpreted as a combinational circuit that is extended by memory elements.

Each circuit design, no matter how complex, has to be verified with respect to numerous aspects before production can take place. Starting from final manufacturing and moving back through previous stages of the design process, a large number of verification aims become obvious. As representatives, we mention the verification of low-level design rules (e.g., layout), timing, high-level design rules (e.g., logic level, RTL level), firmware, and functional correctness. In the following, we restrict our discussion to the problem of *functional correctness* of a circuit design.

In every design step, the exact input/output behavior of the system has to be specified. The designer tries to realize this functional behavior within the bounds of the given technology. The description of the system behavior is called **specification**. A realization of the specified system behavior is called **implementation**. Now the design process of a system can be reduced to an iteration of specification/implementation steps by means of a successive refinement: the implementation of step i becomes the specification of step

$i+1$ (see Fig. 5.1). If it can be *proven* by any method that all implementations satisfy the given specifications, then the chip design is functionally correct.

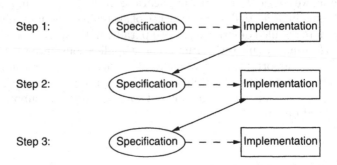

Figure 5.1. Successive refinement of specifications and implementations

For this proof of functional correctness, three concepts are of particular practical interest: *simulation, formal verification,* and *partial verification.*

Simulation means systematically evaluating the functional behavior of both the implementation and the specification for a large number of input vectors. If the output of the implementation and the specification agree for all these vectors, the design is supposed to work correctly. However, the permanently growing complexity of circuit designs renders the complexity of this traditionally applied method increasingly uncontrollable.

The aim of **formal verification** is a *formal mathematical proof* that the functional behavior of the specification and the implementation coincide. The use of ordered binary decision diagrams, called OBDDs, has given this approach quite essential impulses and opened a new age of formal verification. In the following, we will go into this topic in more detail.

Partial verification also uses formal mathematical methods to prove that the implementation at least satisfies particularly important properties of the specified system behavior. These important properties include *safety properties* and *liveness properties.* **Safety properties** express that certain "bad" events cannot occur in any phase of the system's processing. **Liveness properties** express that in any phase of the system's processing, it is possible that certain "good" events eventually occur. Of course, verification of safety and liveness properties cannot provide a proof of the complete functional correctness of the system. It is merely guaranteed that some particularly important properties of the implemented system are satisfied.

5.2 Formal Verification of Combinational Circuits

First, we consider in some detail the verification issue involved in what is called the **logic synthesis** design phase of a combinational circuit. Here, the functionality of the circuit to be designed is given in terms of a network of logical gates. A **gate** consists of a suitable collection of transistors such that the output of the gate computes the result of a certain Boolean operation on its inputs. For example, the AND-gate with two inputs x and y computes the switching function $f(x, y) = x \cdot y$. A central task in logic synthesis is to optimize the representation of this circuit on the gate level with respect to several factors. These optimization criteria include the number of gates, the chip area, energy consumption, delay, clock period, etc. Due to the large complexity of the circuits to be optimized, this task cannot be achieved without the support of sophisticated and comprehensive software, called logic synthesis systems.

In the historical development, several milestones in the history of logic synthesis systems were passed at academic institutions. These software packages substantially influenced the commercial tools of the next generation, or were even adopted in industrial environments. Some of the most important logic synthesis systems in the historical review were:

- MINI (IBM Research, 1974),
- ESPRESSO (University of California at Berkeley, 1984),
- MIS (University of California at Berkeley, 1987),
- BOLD (University of Colorado at Boulder, 1989),
- SIS (University of California at Berkeley, 1992),

as well as the commercial systems of leading design tool companies as Synopsis, Cadence, or Mentor Graphics.

Even without going into the details of the different software systems, it is clear that all the systems have to employ a suitable representation which allows the switching functions to be manipulated. These representations are realized internally by means of suitable data structures. The memory consumption and the running time of individual algorithms within the software systems depend primarily on the chosen representation and the employed data structures. While real memory consumption typically coincides quite exactly with the size of the abstract representation, it is more difficult to determine the efforts of different tasks with respect to time resources.

Before determining the time complexity of the most important routines within logic synthesis systems (which will be done in the next chapters), one has to state which routines are the most important ones. In the following, we state the central problems that have to be solved by any of these systems.

The first of these problems is to generate the internal circuit representation from a given net list of gates. This process is called **symbolic simulation**. Here we have to construct the representations of the literals, i.e., the representations of the functions being computed in the input nodes. Then, in topological order, the representations of the functions being computed in the individual gates are determined depending on the functions of the corresponding predecessor gates. This is possible by applying the Boolean operation performed in the gate to the representations of the inputs of the gate. Figure 5.2 illustrates this process for an AND-gate and the representation type of disjunctive normal forms.

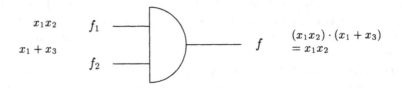

Figure 5.2. Symbolic simulation of an AND-gate in the representation type of a disjunctive normal form

Since, in general, combinational circuits are only available to logic synthesis programs in the form of a net list of gates, high performance of symbolic simulation is absolutely necessary. Primarily, this demand refers to the employed representation of the switching functions. As we have already seen, big differences can be observed among the individual representation types. There are representations where it is almost trivial to perform a Boolean operation, whereas for other representations this task is **NP**-hard and hence hopelessly difficult. Consequently, representation types in circuit design should necessarily satisfy the following property:

(1) Boolean operations can be performed efficiently.

Example 5.6. An important step in logic synthesis is to optimize a circuit on the logic level with regard to the number of gates from a given gate library. We consider the problem of transforming the circuit in Fig. 5.3 (a) to a smaller circuit which only consists of NAND-gates. A typical solution computed by a logic synthesis tool may look like the circuit in Fig. 5.3 (b). It remains to show that the circuits are functionally equivalent, i.e., that they compute the same function. Indeed, in those small dimensions we can convince ourselves even manually. Namely, DeMorgan's rules imply:

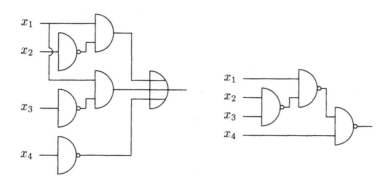

(a) Initial circuit   (b) Realization in terms of NAND-gates

Figure 5.3. Optimizing a circuit by means of NAND-gates

$$\overline{\overline{x_2 \cdot x_3 \cdot x_1} \cdot \overline{x_4}} = \overline{x_2 \cdot x_3} \cdot x_1 + \overline{x_4}$$
$$= (\overline{x_2} + \overline{x_3}) \cdot x_1 + \overline{x_4}$$
$$= x_1 \overline{x_2} + x_1 \overline{x_3} + \overline{x_4}.$$

◇

To guarantee that the functionality of a circuit has not been modified through
the synthesis and optimization process, a verification task has to be solved
as described in Example 5.6. Here, the representation of the original circuit
is interpreted as specification. The representation of the optimized circuit is
considered as implementation. It has to be proven formally that specification and implementation are functionally equivalent, i.e., that both compute
exactly the same switching function. Typically, this proof is carried out in
several steps: alternately, some optimization steps are applied, and then the
resulting partial solutions are verified. As a precondition, it is necessary
that both the optimization steps and the proof of functional equivalence can
be performed efficiently. The first condition can be solved quite well in the
mentioned systems, but it turns out to be difficult to satisfy the second condition. In particular, the exact difficulty of this proof depends directly on the
properties of the representation of switching functions which is used. Hence,
representation types of switching functions in computer-aided circuit design
should have the following property:

(2) Functional equivalence can be tested efficiently.

The equivalence test also plays a central role within the concept of partial
verification. Later on, we will show that efficient testing the functional equivalence of two circuit representations allows us to prove important properties
of a circuit efficiently.

The term **satisfiability test of circuits** denotes the test whether there is an input vector that computes a 1 in the given circuit. The question of how far the satisfiability test can be performed efficiently, has gained a key position in theoretical computer science. In many investigations it serves as a reference problem. If one has access to an efficient test concerning functional equivalence of two circuit representations, then satisfiability can be efficiently checked, too: it is merely to check whether the representation of the contradiction function (which can mostly be obtained in a trivial way) is functionally equivalent to the representation of the given circuit. If the answer is "Yes", then the circuit cannot be satisfied; if the answer is "No", then there exists at least one satisfying assignment.

5.3 Formal Verification of Sequential Circuits

While the outputs of a combinational circuit are completely determined by the current inputs, the outputs of a sequential circuit additionally depend on values computed in the past. The dependency on the past is established by the possibility of storing signals in registers. The outputs of the sequential circuit can depend both on the current inputs and on the values in the memory elements. Therefore, sequential circuits can be seen as an interconnection of combinational logic gates and registers. Figure 5.4 illustrates this relation. Figure 5.5 shows a simple sequential circuit.

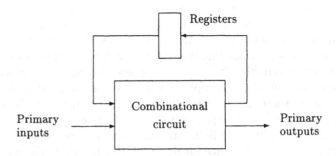

Figure 5.4. Schematic view of a sequential circuit

Whereas the behavior of combinational circuits can be expressed completely in terms of switching functions in a static way, sequential circuits additionally show a dynamic input/output behavior with respect to time. Their behavior can be captured adequately by means of finite state machines, which were introduced in Section 2.7.

Solving the verification problem of sequential circuits is essentially more difficult than solving the one of combinational circuits, as it requires checking

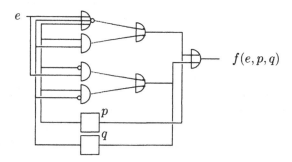

Figure 5.5. Example of a sequential circuit with two registers p, q

permanent functional equivalence for all possible input sequences. In fact, the problem is to show that every input sequence produces the same output sequence in both circuits, where the systems are described as net lists of gates again. Here, too, one system specifies a functional behavior, and the other one an optimized implementation.

The *equivalence test of finite state machines* has been investigated in computer science for many years. For systems whose states are encoded by, say, 80 bits, the number of possible states is 2^{80}. Such a huge number is hardly accessible to intuitive comprehension and may be appreciated by a comparison from the real world. The age of our whole universe is estimated to 2^{34} years, so a computer which had been investigated 2 million states per second since the Big Bang would still not have completed its analysis of all states ! As a consequence of this dilemma, for a long time the correctness of sequential systems in practical applications has only been verified by a large number of simulations. However, even a large number of simulations is small compared to the total number of all states, so that generally, only a small fraction of cases can be covered. In contrast to this, *formal verification* is expected to provide a complete proof of the correctness of the sequential circuit.

Classical approaches for solving the verification problem were doomed to failure for large sequential systems, as they were based on *explicit* representations of state sets, e.g., in the form of lists. Of course, algorithms for state sets in the mentioned sizes cannot be practical. In recent years, fundamental relations between this issue and the representation aspect of switching functions have been revealed, which have recognized the dependency on the algorithmic qualities of the representations. If a representation type is provided which can perform the relevant tasks quickly, then one immediately obtains efficient algorithms for the verification of finite state machines.

The modern approaches are based on an **implicit set representation**. Here, one considers a subset $S \subset \{0, 1\}^n$, which may contain all n-tuples of register assignments that can be reached by arbitrary input sequences of length k. S is a set, but the **characteristic function** χ_S of S,

$$\chi_S(x_1, \dots, x_n) = 1 \iff (x_1, \dots, x_n) \in S,$$

is a switching function in n variables. Hence, set representations and set operations can be completely reduced to the manipulation of switching functions. If a compact representation of switching functions is provided, the implicit set descriptions remain small, too. Even the equivalence test of two finite state machines M_1 and M_2 can itself be reduced to manipulating switching functions by means of implicit set representations.

Indeed, nowadays these implicit techniques are an indispensable tool for solving important problems within the basic model of finite state machines. As an additional requirement on representation types of switching functions in computer-aided circuit design we demand that:

(3) Basic algorithms on finite state machines like the equivalence test can be performed efficiently.

As will be shown later, when reducing these algorithms to the level of switching functions, new basic problems like **quantification** come to the center of attention. These new problems constitute the present challenges with respect to the underlying representation types. With regard to our aim in this chapter, namely describing paradigmatic applications of switching functions in computer-aided circuit design, we can be content with the above formulation.

5.4 References

The mentioned logic synthesis systems are described in the following references: MINI [HCO74], ESPRESSO [BHMS84], MIS [BRSW87], BOLD [HLJ+89] and SIS [SSM+92]. Discussions concerning the requirements on data structures in formal circuit verification can also be found in the surveys [Bry92, Weg94, MT98].

Part II

OBDDs: An Efficient Data Structure

6. OBDDs – Ordered Binary Decision Diagrams

My power comes to full strength in weakness.
2. Corinthians 12,9

In this chapter, we introduce the representation type of *ordered binary decision diagrams*, called *OBDDs* for short. Although the underlying model of decision diagrams (or synonymously branching programs) was already studied by Lee and Akers in the 1950s and 1970s, these representations have not been used in serious applications for a long time. In 1986, by adding some ingenious ordering restrictions to these models and providing a sophisticated reduction mechanism, R. Bryant substantially improved the model. Since this time, the improved representation, denoted as OBDD, has invaded nearly all areas of computer-aided VLSI design.

These facts have been established by the following valuable properties of OBDDs:

- Reduced OBDDs provide a canonical representation of switching functions.
- Reduced OBDDs can be manipulated efficiently.
- For many practically important switching functions, the corresponding OBDD representations are quite small.

Exactly these properties form the basis for efficiently solving the basic problems in computer-aided circuit design that were discussed in Chapter 5.

6.1 Notation and Examples

Definition 6.1. *Let π be a total order on the set of variables $x_1, \ldots, x_n$. An **ordered binary decision diagram** with respect to the variable order π is a directed acyclic graph with exactly one root which satisfies the following properties:*

- *There are exactly two nodes without outgoing edges. These two nodes are labeled by the constants 1 and 0, respectively, and are called **sinks**.*

- *Each non-sink node is labeled by a variable x_i, and has two outgoing edges, which are labeled by 1 and 0, respectively. These edges are called the* **1-edge** *and the* **0-edge**, *respectively.*
- *The order, in which the variables appear on a path in the graph, is consistent with the variable order π, i.e., for each edge leading from a node labeled by x_i to a node labeled by x_j it holds that $x_i <_\pi x_j$.*

Nodes labeled by a variable x_i are called **internal nodes**. *The variable of a node v is abbreviated by $var(v)$. The successor node determined by the 1-edge is denoted by $high(v)$, and the successor node determined by the 0-edge is denoted by $low(v)$.*

In the terminology of the previous chapters, an OBDD is a read-once branching program with an additional ordering restriction on the variables. The **computation path** of an input $a = (a_1, \dots, a_n) \in \mathbb{B}^n$ is the path from the root to a sink in the OBDD which is defined by the input. More precisely, the computation path begins in the root, and in each node labeled by x_i the path follows the edge with label a_i.

Definition 6.2. *An OBDD represents a given switching function $f \in \mathbb{B}_n$ if for all inputs $a \in \mathbb{B}^n$, the computation path of a leads to the sink with label $f(a)$.*

Example 6.3. Let π be the variable order $x_1 < x_2 < x_3$. Figure 6.1 shows two OBDD representations of the function $f(x_1, x_2, x_3) = x_1 x_2 + x_1 \overline{x_2}\ \overline{x_3}$ with respect to the order π. $\diamond$

A node v with label x_i in the OBDD defines a Shannon expansion. If the OBDD rooted in v represents the function $f(x_1, \dots, x_n)$, then the two sub-OBDDs rooted in the sons represent the functions $f(x_1, \dots, x_{i-1}, 1, x_{i+1}, \dots, x_n)$ and $f(x_1, \dots, x_{i-1}, 0, x_{i+1}, \dots, x_n)$, respectively. Figure 6.2 illustrates this relation.

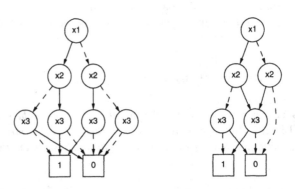

Figure 6.1. Two OBDDs of $f(x_1, x_2, x_3) = x_1 x_2 + x_1 \overline{x_2}\ \overline{x_3}$

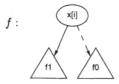

Figure 6.2. Shannon's expansion in the OBDD of f, where $f_1 = f(x_1, \ldots, x_{i-1}, 1, x_{i+1}, \ldots, x_n)$ and $f_0 = f(x_1, \ldots, x_{i-1}, 0, x_{i+1}, \ldots, x_n)$

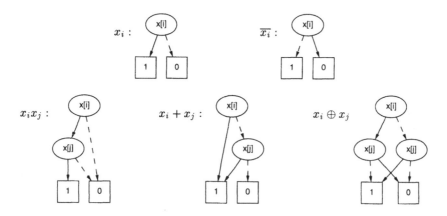

Figure 6.3. OBDDs of elementary functions

Definition 6.4. *Let $f = f(x_1, \ldots, x_n) \in \mathbb{B}_n$. The **positive cofactor** of f with respect to x_i is the subfunction $f_{x_i=1} = f(x_1, \ldots, x_{i-1}, 1, x_{i+1}, \ldots, x_n)$, for short denoted by f_{x_i}. The **negative cofactor** of f with respect to x_i is the subfunction $f_{x_i=0} = f(x_1, \ldots, x_{i-1}, 0, x_{i+1}, \ldots, x_n)$, abbreviated by $f_{\overline{x_i}}$.*

If the root of the OBDD is labeled by the variable x_i, then the Shannon decomposition which is realized in this node can be written in the form

$$f = x_i f_{x_i} + \overline{x_i} f_{\overline{x_i}}.$$

Example 6.5. To complete this introduction of notation, we list some OBDDs of basic functions. Here, we assume that in the variable order the variable x_i occurs before the variable x_j. In Fig. 6.3, the OBDDs of the literals x_i, $\overline{x_i}$, and the functions $x_i x_j$, $x_i + x_j$, $x_i \oplus x_j$ are shown.   ◇

6.2 Reduced OBDDs: A Canonical Representation of Switching Functions

At first, OBDDs in the sense of the stated definitions are not uniquely determined. A corresponding example has already been given in Example 6.3. In his fundamental publication, Bryant investigated the redundancies within decision diagrams. Based on this, he introduced the notion of *reduced* OBDDs which provide a canonical representation of switching functions. In the following, we omit mentioning the variable order π explicitly if it is clear from the context.

One can easily see that the following types of redundancy can occur within an OBDD.

- The 0- and the 1-successor of a node v can be identical. In this case, the decision in the node v does not yield any new information.
- Within the decision diagram, certain subgraphs can occur several times. In this way, the same information about the function is represented several times.

The following definitions make it possible to state the second form of redundancy in a precise manner. It says that two OBDDs can be denoted as isomorphic if and only if they are isomorphic as node- and edge-labeled graphs.

Definition 6.6. *Let P_1 and P_2 be two OBDDs. P_1 and P_2 are called **isomorphic** if there is a bijective mapping ϕ from the set of nodes of P_1 to the set of nodes of P_2 such that, for each node v, either*

1. *the two nodes v and $\phi(v)$ are sinks with identical labels, or*
2. *$var(v) = var(\phi(v))$, $\phi(high(v)) = high(\phi(v))$, $\phi(low(v)) = low(\phi(v))$.*

Definition 6.7. *An OBDD is called **reduced** if*

1. *it does not contain a node v with $high(v) = low(v)$, and*
2. *there does not exist a pair of nodes u, v such that the sub-OBDDs rooted in u and v are isomorphic.*

Note that the second condition for a reduced OBDD refers to a global property of the graph, since the isomorphic subgraphs can be arbitrarily large. However, this global property can be replaced by a local property. Indeed, the local description is better suited for the algorithmic realization of the reduction concept. The local properties are expressed in terms of reduction rules that can be applied to an OBDD until it is completely reduced.

Definition 6.8. *We consider two reduction rules on OBDDs:*

Elimination rule: *If the 1-edge and the 0-edge of a node v point to the same node u, then eliminate v, and redirect all incoming edges of v to u.*

Merging rule: *If the internal nodes u and v are labeled by the same variable, their 1-edges lead to the same node, and their 0-edges lead to the same node, then eliminate one of the two nodes u, v, and redirect all incoming edges of this node to the remaining one.*

The two reduction rules of OBDDs are illustrated in Fig. 6.4.

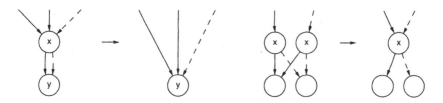

Figure 6.4. The two reduction rules of OBDDs

The following theorem justifies use of the term reduction rules when speaking of the elimination or the merging rule.

Theorem 6.9. *An OBDD is reduced if and only if neither of the two reduction rules can be applied.*

Proof. If an OBDD is reduced, then by definition, the elimination rule cannot be applied. The merging rule cannot be applied either, as otherwise the OBDD would contain two isomorphic subgraphs.

Conversely, let P be an OBDD to which neither of the reduction rules can be applied. Then, by definition of the elimination rule, the OBDD P does not contain a node v with $high(v) = low(v)$.

Now we assume that P contains a pair of different nodes u, v with the property that the subgraphs rooted in u and v are isomorphic.

Case 1: It holds $high(u) = high(v)$ and $low(u) = low(v)$. The merging rule can be applied. This is a contradiction to the precondition that *none* of the reduction rules can be used.

Case 2: It holds $high(u) \neq high(v)$ or $low(u) \neq low(v)$. W.l.o.g. we assume that $high(u) \neq high(v)$. Definition 6.6 implies that the two OBDDs rooted in the two different nodes $high(u)$ and $high(v)$ are isomorphic, too. For this reason, we apply the case distinction once again, this time to the nodes

$high(u)$ and $high(v)$. As these sub-OBDDs only depend on the variables which occur after $var(u)$ in the order, the recursive process stops after at most n steps, where n is the number of variables in the original OBDD. Then, finally, the condition of case 1 has to hold, as there is only one 1-sink and one 0-sink. $\qquad\square$

The statement that iterated application of the reduction rules leads to a reduced OBDD does not tell us yet that the resulting reduced OBDD is uniquely determined. The following theorem establishes one of the key property of OBDDs, namely that reduced OBDDs provide a canonical representation of switching functions: with respect to a fixed variable order, each function can be represented by a uniquely determined OBDD. The proof of this property is carried out in several steps. First we show that the functions which are computed in the nodes of the OBDDs are certain subfunctions of f. By means of these subfunctions, OBDDs of minimal size with regard to f can be described. Based on these steps we can prove that all reduced OBDDs are pairwise isomorphic and that they yield the minimal OBDD of f.

For simplicity, we assume for the rest of this section that the variables $x_1, \dots, x_n$ are naturally ordered, i.e., corresponding to their indices. This is no restriction, as we can obtain this order from every other order by renaming the variables.

Let v be a node labeled by x_i in an OBDD of the function f. Further let p be a path which leads from the root to v. Of course, only the variables x_j with $j \leq i - 1$ may be tested on p. The path p is traversed by an input $(c_1, \dots, c_n)$ if all variables tested on p have suitable values, i.e., if c_j coincides with the value of the edge in p which starts in the node labeled by x_j. In the node v and in the nodes below v only variables $x_i, \dots, x_n$ may be tested. Hence, in the sub-OBDD rooted in v, a subfunction $f_{x_1=c_1,\dots,x_{i-1}=c_{i-1}}$ of f is computed.

If besides p there is another path q leading from the root to v and traversed for the inputs $(d_1, \dots, d_n)$, then in v the function $f_{x_1=d_1,\dots,x_{i-1}=d_{i-1}}$ is computed, too. As in the nodes of the OBDD below v, only the variables $x_i, \dots, x_n$ may be tested, the identity $f_{x_1=c_1,\dots,x_{i-1}=c_{i-1}} = f_{x_1=d_1,\dots,x_{i-1}=d_{i-1}}$ holds. Hence, in each node labeled by x_i of an OBDD, exactly the subfunction of f is computed which we obtain from f by suitably fixing $x_1, \dots, x_{i-1}$ to constants.

Now we can describe OBDDs for any switching function.

Theorem 6.10. *Let S_j be the set of subfunctions of f which we obtain by fixing $x_1, \dots, x_{j-1}$ to constants, and which depend essentially on x_j. Up to isomorphisms, there is exactly one OBDD of minimal size for f with respect to the variable order $x_1, \dots, x_n$. This OBDD contains exactly $|S_j|$ nodes labeled by x_j.*

Proof. We explicitly construct a minimal OBDD P of f which for all i contains exactly $|S_i|$ nodes labeled by x_i. If f is a constant function, then P exactly consists of the two sinks 0 and 1. Otherwise, P contains a 1-sink, a 0-sink, and for each subfunction $g \in S_i$ exactly one node v which is labeled by x_i. Let g_{x_i} and $g_{\overline{x_i}}$ be the positive and negative cofactor of g. For these subfunctions, P contains the nodes v_1 and v_0. Each of these nodes is either a sink (if g_{x_i} or $g_{\overline{x_i}}$ is a constant function), or an internal node labeled by x_j where $j > i$. We choose v_1 to be the 1-successor of v, and v_0 to be the 0-successor of v.

In order to prove that the constructed OBDD P computes the function f, it suffices to prove that at each node v the corresponding subfunction is computed. We prove this statement by induction where we traverse the nodes in a reverse topological order. First, the statement holds for the sinks, as the sinks compute the constant functions. For an internal node v labeled by x_i which computes the function g, the induction hypothesis implies that the 1-successor computes the positive cofactor of g, and the 0-successor computes the negative cofactor of g with respect to x_i. Hence, in v the function $g = x_i\, g_{x_i} + \overline{x_i}\, g_{\overline{x_i}}$ is computed.

Minimality of P: If the constructed OBDD P is not of minimal size, then there exists an OBDD P' which contains for some $i \in \{1, \ldots, n\}$ less than $|S_i|$ nodes labeled by x_i. However, as there are $|S_i|$ different subfunctions of f of the form

$$f_{x_1 = c_1, \ldots, x_{i-1} = c_{i-1}}$$

which depend essentially on x_i, there exists an assignment $c = (c_1, \ldots, c_{i-1})$ to the variables $x_1, \ldots, x_{i-1}$ with $f_{x_1 = c_1, \ldots, x_{i-1} = c_{i-1}} \in S_i$, whose corresponding path leads to either a sink, or to a node labeled by x_j, $j > i$, or to a node labeled by x_i that computes a different subfunction $f_{x_1 = d_1, \ldots, x_{i-1} = d_{i-1}}$. The first two cases lead to contradictions, as $f_{x_1 = c_1, \ldots, x_{i-1} = c_{i-1}}$ depends essentially on x_i, and the last case leads to a contradiction, as in each node of an OBDD only one subfunction can be computed.

Minimal OBDDs are isomorphic to P: As a consequence, *each* minimal OBDD of f with respect to the variable order $x_1, \ldots, x_n$ has to contain the following nodes: for each subfunction $g \in S_i$, there has to be a node labeled by x_i such that the successors of the node compute the cofactors g_{x_i} and $g_{\overline{x_i}}$. By combining these subgraphs we obtain: each OBDD of f which contains exactly the same number of nodes as the constructed OBDD P is isomorphic to P. Hence, an OBDD of f which is not isomorphic to P has to contain additional nodes, and therefore it is not minimal.   □

Theorem 6.11. *Let P be an OBDD of the switching function f with respect to the variable order π. P is isomorphic to the minimal OBDD P' of f with respect to π if and only if none of the reduction rules is applicable to P.*

Proof. Obviously, neither of the reduction rules can be applied to a minimal OBDD; hence, it suffices to proof that in case of an OBDD P which is not minimal at least one of the rules can be applied.

W.l.o.g. we can assume that f is not a constant function. Moreover, for the ease of notation we assume that the natural variable order $x_1, \ldots, x_n$ is used. The proof of Theorem 6.10 implies that each OBDD of a switching function f contains for each subfunction in S_i at least one node labeled by x_i. If P is not isomorphic to the minimal OBDD P', then there is at least one $i \in \{1, \ldots, n\}$ such that P contains more than $|S_i|$ nodes labeled by x_i. Let i' be the greatest of these i. Then, in P, each of the subfunctions in S_j with $j > i'$, and each of the two constant functions is represented by exactly one node. As there are more than $|S_{i'}|$ nodes labeled by $x_{i'}$, P contains either a node u labeled by $x_{i'}$ which computes a subfunction g that does not depend essentially on $x_{i'}$, or there exist two nodes v and w which are labeled by $x_{i'}$ and which compute the same subfunction $h \in S_{i'}$. In the first case we have $g = g_{x_{i'}} = g_{\overline{x_{i'}}}$. As the functions $g_{x_{i'}}$ and $g_{\overline{x_{i'}}}$ are represented in the same node, the node u can be removed by means of the elimination rule. In the second case the successors of v and w compute the functions $h_{x_{i'}}$ and $h_{\overline{x_{i'}}}$. For each of these functions there is only one node in P, and hence, v and w can be identified by means of the merging rule. $\square$

Corollary 6.12. *For each variable order π, the reduced OBDD of a switching function f with respect to π is uniquely determined (up to isomorphisms).*

Proof. The statement follows immediately from Theorem 6.11 in connection with Theorem 6.9. $\square$

To complete this section, let us remark that all the OBDDs of the basic functions given in Example 6.5 are reduced.

6.3 The Reduction Algorithm

The reduction rules of Definition 6.8 in connection with Theorem 6.11 suggest an easy algorithm for transforming a given OBDD into a reduced OBDD of the same function: apply the rules as long as this is possible. Each application of one of the two rules decreases the size of the OBDD by at least one node. If no reduction rule can be applied any longer, then the OBDD is reduced.

The algorithm in Fig. 6.5 performs the reduction steps in a systematic and efficient manner. The main idea of the algorithm has already been employed in Theorem 6.11: in order to reduce a given OBDD it is most reasonable to proceed bottom-up. Otherwise, it cannot be guaranteed that the execution of a reduction does not make a new reduction possible, within an already investigated area. For this reason, first the applicability of a reduction rule

```
Reduce(P) {
/* Input: An OBDD P of f(x₁,...,xₙ) ∈ 𝔹ₙ
              with respect to the natural variable order */
/* Output: A reduced OBDD of f */
    Assign to each node v a positive number id(v) in a bijective way;
    For i = n,...,1 (decreasing) {
          V(i) = {v node in P : var(v) = xᵢ};
          /* Elimination rule */
          For all v ∈ V(i) {
                If id(low(v)) = id(high(v)) {
                      Remove v from V(i);
                      Redirect all incoming edges of v to low(v);
                      Remove v;
                }
                Else {
                      key(v) = (id(low(v)), id(high(v)));
                }
          }
          /* Merging rule */
          oldkey = (0,0);
          For all v ∈ V(i), sorted by key(v) {
                If key(v) = oldkey {
                      Remove v from V(i);
                      Redirect all incoming edges of v to oldnode;
                      Remove v;
                }
                Else {
                      oldnode = v;
                      oldkey = key(v);
                }
          }
    }
}
```

Figure 6.5. Reduction algorithm

is checked for those nodes that are labeled by the last variable in the order. Subsequently, the next variable, i.e., the one immediately preceding in the order, is investigated.

At the beginning of the algorithm each node v is mapped bijectively to a positive number $id(v)$ which serves as identification. The test if the elimination rule is applicable can be carried out for each node locally. In order to check whether the merging rule can be applied, it seems to be a good strategy to sort the nodes v of the currently investigated variable according to the key $(id(low(v)), id(high(v)))$. The result of this sorting process is that nodes which can be identified by means of the merging rule now appear consecutively.

The time complexity of the algorithm is dominated by the time for sorting the subsets of nodes. If the OBDD P consists of m nodes, then the running time of the method is bounded from above by $\mathcal{O}(m \log m)$. By using the so-called bucket sort technique the method can even be modified in a way that it merely requires linear time. However, from a practical point of view this variant has two drawbacks. First, the constants in the asymptotic $\mathcal{O}$-Notation and therefore the running times are quite large, and second, a large amount of additional memory is required.

Example 6.13. Figure 6.6 shows an OBDD whose nodes have already been assigned positive numbers in a bijective manner. The reduction algorithm works on these OBDDs as follows. In the first traversal of the loop body it is recognized that the node with label x_3 has identical 1- and 0-successors. Hence, it can be removed. In the second traversal of the loop, the merging rule is applied to the two nodes with label x_2. Finally, in the last step the node with label x_i is removed by means of the elimination rule.   ◇

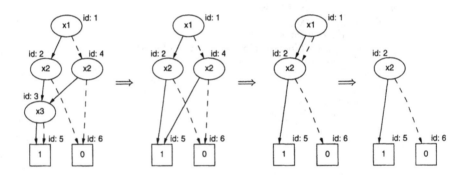

Figure 6.6. Reduction of an OBDD

6.4 Basic Constructions

In this section, we describe two fundamental techniques for constructing reduced OBDDs of given functions. These two techniques are based on different approaches. In particular, both approaches are suitable for manually constructing OBDDs of functions in few variables. Here, for example, the starting point may be a formula representation of a function. Indeed, we recommend that the reader constructs some OBDDs of simple functions manually, as this serves to provide quite good insights into the character and shape of OBDD representations.

The first method is based on the reduction algorithm presented in Section 6.3. Let $f \in \mathbb{B}_n$, and let π be a given order on the variables. In order to construct the reduced OBDD of f, one starts by constructing a complete decision tree of f that reads the variables on each path according to the order π. Of course, the construction of the decision tree requires determination of all 2^n function values of f.

By identifying all 1-sinks and all 0-sinks, respectively, one obtains an OBDD of f with respect to the variable order π. Finally, the reduction algorithm is used to reduce this OBDD. Remember that the reduction algorithms reduces the OBDD in bottom-up manner, i.e., from the sinks to the root. Of course, the need to construct a complete decision tree in the initial step makes the approach practical only for functions in quite few variables.

The second method constructs the reduced OBDD in a top-down manner, i.e., from the root to the sinks. First, the root node of the reduced OBDD is introduced. This node is labeled by the first variable x_i in the order which is essential for the function f. By means of Shannon's decomposition

$$f = x_i f_{x_i} + \overline{x_i} f_{\overline{x_i}},$$

we can construct the two subfunctions f_{x_i} and $f_{\overline{x_i}}$ which are computed in the sons of the root node. For each of these functions f_{x_i} and $f_{\overline{x_i}}$ we also determine the first variable in the order which is essential for the function. In either case this variable becomes the label of the corresponding node. Recursively, this method is continued. The procedure for determining the node labels serves to completely prevent constructing redundant nodes. If one additionally takes care that already represented subfunctions will not be represented a second time, then the merging rule is also satisfied. The whole process is finished when constant subfunctions are reached. Of course, this approach is practical only for small n, too, as the test whether a certain subfunction has already been represented requires an individual equivalence test.

Example 6.14. Let $f(x_1, x_2, x_3, x_4) = x_2(x_3 + \overline{x_4}) + \overline{x_1}\ \overline{x_2}x_4 + x_1\overline{x_2}\ \overline{x_4}$, and let π be the variable order $x_1 < x_2 < x_3 < x_4$. Figure 6.7 shows the order in which the nodes of the OBDDs are constructed in the presented top-down approach. First the root node is constructed. The two subfunctions $f_{x_1} = f_{x_1=1}$ and $f_{\overline{x_1}} = f_{x_1=0}$ are

$$f_{x_1} = \overline{x_2}\ \overline{x_4} + x_2(x_3 + \overline{x_4}),$$
$$f_{\overline{x_1}} = \overline{x_2}x_4 + x_2(x_3 + \overline{x_4}).$$

The functions have not been represented in the OBDD yet. As both of them essentially depend on x_2, we add two nodes which are labeled by x_2. In the next step we compute the cofactors with respect to x_2,

$$f_{x_1 x_2} = f_{\overline{x_1} x_2} = x_3 + \overline{x_4},$$
$$f_{x_1 \overline{x_2}} = \overline{x_4},$$
$$f_{\overline{x_1}\ \overline{x_2}} = x_4.$$

This leads to three functions which have not been represented yet (nodes 4, 5, 6), and one of them essentially depends on x_3. For these functions, the cofactors with respect to x_3 will be computed:

$$f_{x_1 x_2 x_3} = f_{\overline{x_1} x_2 x_3} = 1,$$
$$f_{x_1 x_2 \overline{x_3}} = f_{\overline{x_1} x_2 \overline{x_3}} = f_{x_1 \overline{x_2} x_3} = f_{x_1 \overline{x_2}\ \overline{x_3}} = \overline{x_4},$$
$$f_{\overline{x_1}\ \overline{x_2} x_3} = f_{\overline{x_1}\ \overline{x_2}\ \overline{x_3}} = x_4.$$

The only one of these functions which has not been represented yet is the constant function 1. Finally, in the last step all resulting subfunctions are constant functions.   ◇

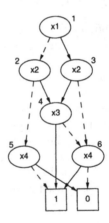

Figure 6.7. Top-down construction of a reduced OBDD

6.5 Performing Binary Operations and the Equivalence Test

Next, we show how two of the basic problems which have been pointed out in Section 5.2 can be solved efficiently if switching functions are represented in terms of OBDDs. The resulting algorithms will constitute the foundation for efficiently employing the OBDD data structure in all applications. By means

```
Apply(F, G, *) {
/* Input: OBDDs F, G of f, g with respect to π; a binary operation * */
/* Output: An OBDD of f * g */
    If (F and G are sinks) {
        Return (F * G);
    }
    Else If (F, G) ∈ IdealComputedTable {
        Return IdealComputedTable(F, G);
    }
    Else {
        Let xᵢ be the first variable in π which is essential for F or G;
        Construct a new node v;
        high(v) = Apply(F_{xᵢ=1}, G_{xᵢ=1}, *);
        low(v) = Apply(F_{xᵢ=0}, G_{xᵢ=0}, *);
        Insert_ideal_computed_table(F, G, v);
        Return v;
    }
}
```

Figure 6.8. OBDD-based application of a binary operation

of clever implementation techniques, which we discuss in the next chapter, the presented algorithms can be further improved.

First we turn to the problem of performing binary operations. Let $*$ denote an arbitrary Boolean operation, e.g., the conjunction or disjunction. To compute the OBDD representation of the function $f * g$ from the OBDDs of the functions f and g, Shannon's expansion with respect to the leading variable x_i in the order can be used:

$$f * g = x_i \left(f|_{x_i=1} * g|_{x_i=1} \right) + \overline{x_i} \left(f|_{x_i=0} * g|_{x_i=0} \right).$$

By recursively constructing the OBDDs P_1 and P_0 of the two functions $f|_{x_i=1} * g|_{x_i=1}$ and $f|_{x_i=0} * g|_{x_i=0}$, an OBDD of $f * g$ can simply be computed by introducing a new node labeled by x_i whose 1-edge points to the root of P_1, and whose 0-edge points to the root of P_0.

However, there is a hitch to this recursive method for computing $f * g$. If one actually performs all the decompositions explicitly, then one has to deal finally with up to 2^n subproblems.

To make the application of binary operations efficient, one employs the following considerations. Each recursive call of the computation procedure has two arguments, which we denote by f^* and g^*. In each call, f^* is a subfunction of f, and g^* is a subfunction of g. Each subfunction of f and g corresponds to exactly one OBDD node. Multiple calls with the same argument pair can be avoided by recalling all the already computed results from a table. In this way, the originally exponential number of decompositions can now be bounded by the product of the two OBDD sizes.

This idea can be realized by means of the pseudo code in Fig. 6.8. First it is checked whether the present case is a terminal case, i.e., if both OBDDs consist of exactly one sink. If this is true, then the result can be easily determined by applying the binary operation to the values of these sinks. The mentioned table for storing already computed intermediate results is realized by means of the table *IdealComputedTable*. If for a pair of nodes the result is not known yet, then it is computed by two recursive calls.

Example 6.15. Let F and G be two OBDDs of the functions $f(x_1, x_2) = x_1 x_2$, $g(x_1, x_2) = \overline{x_1}\ \overline{x_2}$ with respect to the variable order $x_1 < x_2$. The nodes in the OBDD of $f + g$ result from the Cartesian product of the nodes of f and the nodes of G. In Fig. 6.9, the nodes of F and G are numbered consecutively. The (two-component) numbering of the resulting OBDD reflects the correspondences among the nodes which are involved in the construction. ◇

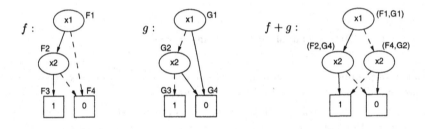

Figure 6.9. Cartesian product involved in binary operations

In general, the algorithm may produce an OBDD which is not reduced. The reduced form can be re-established by applying the already presented (linear) reduction algorithm to the constructed OBDD. Another efficient possibility is to take care that at each moment within the construction the OBDD is kept in a reduced form. This technique will be extensively discussed in the next chapter. First we summarize:

Theorem 6.16. *Let the two switching functions f_1 and f_2 be represented by the OBDDs P_1, P_2 with respect to the same order π. For each binary operation $*$, the reduced OBDD P of the function $f = f_1 * f_2$ with respect to π can be constructed in time $\mathcal{O}(size(P_1) \cdot size(P_2))$.* □

The test whether two given OBDDs P_1 and P_2 represent the same function can also be performed by using the uniqueness of the representation. First, P_1 and P_2 will be reduced. As the reduced OBDD of each function is uniquely determined up to isomorphisms, it is merely to check whether the two OBDDs are isomorphic, i.e., whether the underlying node- and edge-labeled graphs are isomorphic. For this, the two OBDDs P_1 and P_2 are simultaneously traversed from the corresponding root by using depth first search. In each step of this depth first search, we pass over either to the 1-successors of the two current nodes or to the 0-successors. For each visited pair of nodes we check whether both nodes have the same label. If this is true for all nodes, then P_1 and P_2 represent the same function. Thus we have:

Theorem 6.17. *Let the switching functions f_1 and f_2 be represented by OBDDs P_1, P_2 with respect to the same order π. The equivalence test of two representations can be performed in time $\mathcal{O}(size(P_1) + size(P_2))$.*   □

The efficiency of the equivalence test can also be improved by suitable implementation techniques. We will come back to this aspect in the next chapter.

6.6 References

The fundamentals of OBDDs, i.e., the model, the reduction idea, efficient algorithms for performing Boolean operations on them, and the equivalence test, go back to Bryant [Bry86, Bry92]. The presented uniqueness theorem follows the presentation of Sieling and Wegener [SW93a, Sie94]. The linear-time reduction algorithm is also due to Sieling and Wegener [SW93b].

7. Efficient Implementation of OBDDs

Alles, was ist, ist vernünftig.
[All, that is, is reasonable.]
Georg Wilhelm Friedrich Hegel (1770–1831)

So far, we have mainly considered structural properties of ordered binary decision diagrams. Several of these properties, like the linear-time equivalence test, already indicate the suitability of OBDDs in the context of Boolean manipulation. However, as the efficiency of all OBDD-based applications depends almost exclusively on the efficiency of the basic OBDD operations, the demands on the performance of these operations are very high. Hence, much research effort has been spent on transforming the basic OBDD concept into fast and memory-efficient implementations.

A key step in this development was achieved by K. Brace, R. Rudell, and R. Bryant who presented a general framework for implementing OBDD packages and developed the first efficient general-purpose package for manipulating OBDDs. During this process they had to make a number of design decisions. The suitability of their decisions is confirmed by the fact that many of the newer packages are still based on the programming techniques they proposed.

In this chapter, we discuss the key concepts of this framework and then present some well-known software implementations of these ideas.

7.1 Key Ideas

The core elements of the data structure that represents a single node in an OBDD are shown in Fig. 7.1. Here, Index is the index i of the variable x_i. High is a pointer to the 1-successor, and Low is a pointer to the 0-successor of the node. The chosen memory sizes of the components allow up to 65536 different variables to be generated. A word is the machine-dependent memory unit that suffices to address the whole virtual address space. In case of a 32-bit architecture, a word consists of 32 bit = 4 byte, and the virtual address space is of size 2^{32}.

Component	Size
Index	2 byte
High	1 word
Low	1 word

Figure 7.1. Core of the record representing a node

While discussing the key ideas of the implementation, we will see that for efficiency reasons it is inevitable to extend this basic record. By storing additional information the performance of individual OBDD algorithms can be substantially improved. However, as practical applications involve large OBDDs with millions of nodes, the inclusion of additional information has to be carried out with great care. Otherwise, the limited memory resources may not suffice for sophisticated and large applications. The art of finding a good compromise goes back to the ingenious work of those researchers who created successful OBDD packages.

7.1.1 Shared OBDDs

Several functions can be represented within a single directed acyclic graph with several roots. In this way, subgraphs which occur in several OBDDs only need to be represented once. This kind of representation is called a **shared OBDD**.

Example 7.1. Figure 7.2 shows a shared OBDD of the switching function $f_1 = (x_1 \equiv x_2)$, $f_2 = \overline{x_2}$, $f_3 = x_1\overline{x_2}$. ◇

This sharing of subgraphs saves much time and space compared to having separate OBDDs. If it can be achieved that each relevant subfunction is

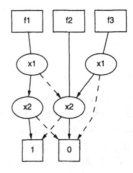

Figure 7.2. Shared OBDD of the function f_1, f_2, f_3

represented exactly once in the shared OBDD, then the canonical representation of OBDDs is even turned into a **strongly canonical form**. This term expresses that two equivalent functions f and g do not only have the same reduced OBDD representation, but that in an implementation they are represented by a pointer to exactly the same memory cell. Consequently, checking two functions f and g for equivalence can be performed immediately by just comparing the root nodes of the OBDDs that correspond to f and g – the equivalence test now only consists of a single pointer comparison !

7.1.2 Unique Table and Strong Canonicity

The strongly canonical form which is possible in connection with shared OBDDs makes it particularly desirable to keep all OBDDs occurring at each moment of the computation in a reduced form. In other words, each relevant subfunction should be represented exactly once.

Whenever a new node has to be added to the shared OBDD, its label x_i, the (already existing) 1-successor *high* and the (already existing) 0-successor *low* have to be specified. To maintain the OBDD in a reduced form, it is checked first whether there already exists a node with this specification. If so, a new node will not be constructed, but the existing node will be used.

The decision whether a triple $(x_i, high, low)$ has already been represented by an OBDD node v is performed by a **unique table**. In order to determine the matching node v from the information $(x_i, high, low)$, it is advisable to implement the unique table by means of a hash table. The triple $(x_i, high, low)$ of a node v is mapped to a hash value $h(x, high, low)$. At this position in the array *uniquetable*, a pointer to the node v is stored. Of course, it is possible that different triples hash to the same value. Therefore, the array *uniquetable* is implemented as an array of linked lists, called collision lists. If the array is large enough and the hash function is well chosen, then each list contains only few elements. The node corresponding to a given triple can then be found very fast.

Example 7.2. In Fig. 7.3, an OBDD of the function $f(x_1, x_2, x_3)$ with respect to the order $x_1 < x_2 < x_3$ is depicted. We denote the internal nodes in the OBDD by $v_1, \ldots, v_5$. The names v_i only serve to distinguish the nodes and could also be the memory addresses of the nodes.

A unique table of this reduced OBDD could be an array of size 5 with indices 0 to 4. A possible hash function for a triple (x_i, v_j, v_k) might be

$$h(x_i, v_j, v_k) = i + j + k \pmod 5.$$

For example, the node v_3 has the hash value $2 + 5 + 7 \pmod 5 = 4$. The unique table of the internal nodes in the OBDD is also illustrated in Fig. 7.3.

$\diamond$

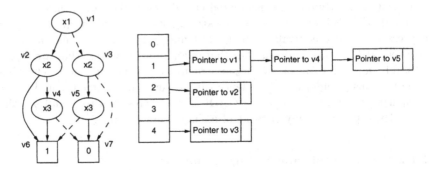

Figure 7.3. Example of a unique table

If the OBDD has been reduced before the lookup, then the following holds. The subfunction which is represented by the triple $(x_i, high, low)$ already exists in the OBDD if and only if there is a node in the list which represents the same triple. In this way, it is easy to maintain the reduced structure of the OBDD while inserting a new subfunction. The unique table works in the same way for OBDDs and shared OBDDs.

An improvement of this concept is to store *the nodes* v_i directly in the lists instead of the pointers the nodes. In this case, the basic record in Fig. 7.1 has to be extended by a component Next containing a reference to its successor node in the corresponding collision list.

7.1.3 ITE Algorithm and Computed Table

In Section 6.5, we have already mentioned the basic idea of computing binary operations in terms of OBDD representations. In order to compute the OBDD of $f * g$ from the OBDDs of two functions f and g for a given operation $*$, Shannon's expansion with respect to a variable x_i can be used:

$$f * g = x_i \left(f|_{x_i=1} * g|_{x_i=1} \right) + \overline{x_i} \left(f|_{x_i=0} * g|_{x_i=0} \right).$$

Recursively, the binary operation with regard to the cofactors is computed.

In order to treat all binary operations in a unified way, Brace, Rudell, and Bryant invented the so-called **If-Then-Else-operator (ITE)**. ITE is a ternary Boolean function with inputs x, y, z that computes the following function: *If x, then y, else z*. Formally, ITE is defined by

$$\text{ITE}(x, y, z) = x \cdot y + \overline{x} \cdot z.$$

The ITE operator is particularly suited in the context of OBDDs, as it reflects Shannon's decomposition performed in a node of the OBDD.

According to Section 3.3.1 there are exactly 16 possible binary operations. As shown in Fig. 7.4, all these 16 operations can simply be expressed in terms of the ITE operation.

No.	Function	ITE operator	No.	Function	ITE operator
0	0	0	8	$f \mathbin{\overline{+}} g$	$\text{ITE}(f,0,\overline{g})$
1	$f \cdot g$	$\text{ITE}(f,g,0)$	9	$f \equiv g$	$\text{ITE}(f,g,\overline{g})$
2	$f \mathbin{\not\Rightarrow} g$	$\text{ITE}(f,\overline{g},0)$	10	$\overline{g}$	$\text{ITE}(g,0,1)$
3	f	f	11	$f \Leftarrow g$	$\text{ITE}(f,1,\overline{g})$
4	$f \mathbin{\not\Leftarrow} g$	$\text{ITE}(f,0,g)$	12	$\overline{f}$	$\text{ITE}(f,0,1)$
5	g	g	13	$f \Rightarrow g$	$\text{ITE}(f,g,1)$
6	$f \oplus g$	$\text{ITE}(f,\overline{g},g)$	14	$f \mathbin{\overline{\cdot}} g$	$\text{ITE}(f,\overline{g},0)$
7	$f + g$	$\text{ITE}(f,1,g)$	15	1	1

Figure 7.4. Realization of the 16 binary operations in terms of the ITE operator

Now let f, g, and h be switching functions, and let x_i be the leading variable in the order. For the computation of $\text{ITE}(f,g,h)$ the following recursive decomposition can be used.

$$
\begin{aligned}
\text{ITE}(f,g,h) &= f \cdot g + \overline{f} \cdot h \\
&= x_i \cdot (f \cdot g + \overline{f} \cdot h)_{x_i} + \overline{x_i} \cdot (f \cdot g + \overline{f} \cdot h)_{\overline{x_i}} \\
&= x_i \cdot (f_{x_i} \cdot g_{x_i} + \overline{f}_{x_i} \cdot h_{x_i}) + \overline{x_i} \cdot (f_{\overline{x_i}} \cdot g_{\overline{x_i}} + \overline{f}_{\overline{x_i}} \cdot h_{\overline{x_i}}) \\
&= \text{ITE}\,(x_i, \text{ITE}(f_{x_i}, g_{x_i}, h_{x_i}), \text{ITE}(f_{\overline{x_i}}, g_{\overline{x_i}}, h_{\overline{x_i}})) \\
&= (x_i, \text{ITE}(f_{x_i}, g_{x_i}, h_{x_i}), \text{ITE}(f_{\overline{x_i}}, g_{\overline{x_i}}, h_{\overline{x_i}}))\,.
\end{aligned}
$$

The triple which has been invented in the last step of this equation has exactly the meaning introduced in the previous section. A node with label x_i and the recursively defined successors has to be constructed if such a node does not exist yet.

The recursion of $\text{ITE}(f,g,h)$ stops if the first argument is constant:

$$
\begin{aligned}
\text{ITE}(1,f,g) &= f, \\
\text{ITE}(0,f,g) &= g.
\end{aligned}
$$

Furthermore the algorithm can already stop in the following cases:

$$
\begin{aligned}
\text{ITE}(f,1,0) &= f, \\
\text{ITE}(f,g,g) &= g.
\end{aligned}
$$

Figure 7.5 shows pseudo code for the ITE algorithm. As in the description of the basic variant of the algorithm in Fig. 6.8, a computed table is employed for storing already computed results. Additionally, the two functions

```
ITE(F, G, H) {
/* Input: OBDDs F, G, H of f, g, h with respect to π */
/* Output: An OBDD of ITE(f, g, h) */
    If (terminal case) {
        Return (result of the terminal case);
    }
    Else If (F, G, H) ∈ ComputedTable {
        Return ComputedTable(F, G, H);
    }
    Else {
        Let xᵢ be the first variable in π which is essential for F, G or H;
        T = ITE(F_{x_i}, G_{x_i}, H_{x_i});
        E = ITE(F_{\overline{x_i}}, G_{\overline{x_i}}, H_{\overline{x_i}});
        R = Find_or_add_unique_table(v, T, E);
        Insert_computed_table({F, G, H}, R);
        Return R;
    }
}
```

Figure 7.5. ITE algorithm

Find_or_add_unique_table and *Insert_computed_table* are used. The first of these functions checks whether a given triple has already been realized by a node in the OBDD. In the positive case it returns a reference to this node, in the negative case the node is newly constructed, and a reference to it is returned. *Insert_computed_table* inserts the currently computed subproblem into the computed table.

Let us consider three OBDDs F, G, H, and all triples of nodes where the first component is a node in F, the second component is a node in G, and the third component is a node in H. If in the computed table all previously computed subresults are stored, then for each of these triples of nodes the ITE algorithm is called at most once. Under the assumption that a look-up and an insert step in the unique table and the computed table requires constant time, then the time complexity of the ITE algorithm is bounded by $\mathcal{O}(size(F) \cdot size(G) \cdot size(H))$.

In case of binary operations, the running time is not only bounded by a cubic function, but even by a quadratic function. Namely, if one of the three ITE arguments is constant and only passed through in each recursion, then ITE has at most quadratic complexity, as only node pairs have to be counted. The complexity of ITE is also quadratically bounded if two of the arguments coincide in each recursive step. Additionally, in the next section we present a technique that allows the complexity of ITE to be bounded by a quadratic function if two of the arguments are complementary.

Of course, these complexity estimates only yield a worst-case estimate. In most practical applications, the time behavior of the ITE algorithm is strongly correlated with the size of the resulting OBDDs.

Example 7.3. Let $f = x_1 + x_2$, $g = x_1 \cdot x_3$, and $h = x_2 \cdot x_4$ be switching functions with the OBDDs F, G, H. The internal nodes in the OBDDs are denoted by B, C, and D as in Fig. 7.6. The OBDD I computed by the function call $\text{ITE}(F, G, H)$ results as follows:

$$
\begin{aligned}
I &= \text{ITE}(F, G, H) \\
&= (x_1, \text{ITE}(F_{x_1}, G_{x_1}, H_{x_1}), \text{ITE}(F_{\overline{x_1}}, G_{\overline{x_1}}, H_{\overline{x_1}})) \\
&= (x_1, \text{ITE}(1, C, H), \text{ITE}(B, 0, H)) \\
&= (x_1, C, (x_2, \text{ITE}(B_{x_2}, 0_{x_2}, H_{x_2}), \text{ITE}(B_{\overline{x_2}}, 0_{\overline{x_2}}, H_{\overline{x_2}}))) \\
&= (x_1, C, (x_2, \text{ITE}(1, 0, 1), \text{ITE}(0, 0, D))) \\
&= (x_1, C, (x_2, 0, D)).
\end{aligned}
$$

The sub-OBDDs C and D already exist in the shared OBDD. Hence, the computation of I is finished, and I has the structure depicted in Fig. 7.6. ◊

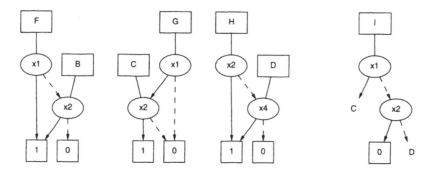

Figure 7.6. Example of computing $I = \text{ITE}(F, G, H)$

In the previous complexity considerations for binary operations, we focused at the running time. Of course, this is justified, as an upper bound for the time complexity also implies an upper bound for the space complexity. However, in case of the ITE algorithm the situation should be investigated in more detail. For two large OBDDs P_1 and P_2 with each more than 100 000 nodes it is surely not possible to provide a computed table with $size(P_1) \cdot size(P_2)$ entries. For this reason, it is reasonable to realize not only the unique table in form of a hash table, but also the computed table. This technique allows fast access without the need to allocate too much memory in advance.

In case of the computed table, however, node triples which map to the same hash value in the ITE algorithm are typically not linked by means of a collision list. Instead of this, the computed table is implemented as a **cache-based hash table**. This term expresses that at each possible function value of the chosen hash function at most k nodes can be stored. If another node is to be stored at an address which already contains k entries, then one of the old entries is replaced.

This type of hash table requires less memory, as it is not necessary to keep all elements in collision lists. Of course, this implementation of the computed table makes it possible that already computed results will be "forgotten" and that they have to be recomputed. As a consequence, the running time analysis for the case of a non-forgetting computing table cannot be applied any longer. In the worst case, all hash values determined in an ITE operation are identical, and the time behavior is exponential. However, when choosing a suitable hash function, extreme cases like this are very seldom.

Besides smaller memory requirements, it is an additional advantage of using a cache-based hash table that *locality* in referencing is exploited. When the ITE algorithm is performed, it occurs quite often that a result is needed again shortly after its initial computation. The longer the time after the computation of a specific result, the smaller becomes the probability that exactly this result is needed again.

If the computed table is realized by a hash table with collision lists, then for space reasons, it is unavoidable to remove the old entries from time to time. In case of a cache-based table this collection process is not necessary, as the old entries are automatically replaced. In experimental studies, a cache depth of 2 has turned out to be favorable, i.e., for each possible function value of the hash function there are two memory cells at which nodes can be stored.

7.1.4 Complemented Edges

A very effective feature implemented in most OBDD packages is the use of **complemented edges**. This technique is based on the fact that the OBDDs of a function f and its complement $\overline{f}$ only differ in one aspect: the values of their sinks are interchanged. By introducing an additional edge attribute of only one bit for each edge, this similarity can be exploited. If the attribute bit is not set, the sub-OBDD which the edge points to is interpreted in the original way as a switching function f. If instead the bit is set, then this sub-OBDD is interpreted as the complement $\overline{f}$ of the ordinary subfunction f. Therefore, the additional edge attribute is also called a **complement bit**. By means of this technique the functions f and $\overline{f}$ can be represented by essentially the same graph: $\overline{f}$ is simply expressed by a complemented edge to the root of the OBDD of f. In this way, a substantial number of nodes can be gained.

In the presence of complemented edges, only one sink is required, as the 0-sink can be represented by the complement of the 1-sink. In our diagrams the edges whose complement bit is set are represented by dotted arrows.

Example 7.4. Figure 7.7 shows the effect of complemented edges when representing two functions f and $\overline{f}$. The complemented edges are drawn as dotted arrows. ◇

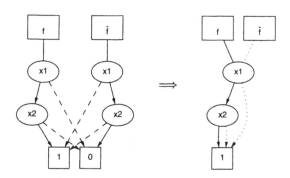

Figure 7.7. Example of complemented edges

The main problem one has to face in the presence of complemented edges is the loss of canonicity in the representation. One way to re-establish this property is to restrict the admissible positions of complement bits. Indeed, some constellations of possible positions of complemented edges are functionally equivalent. If we consider for a node v the triple of the outgoing 1-edge, the outgoing 0-edge, and the incoming edge, there are exactly 8 possible ways to place the complement bits. The following equation, based on DeMorgan's rules, implies that some of these possible ways are functionally equivalent:

$$
\begin{aligned}
\overline{x_i f_{x_i} + \overline{x_i} f_{\overline{x_i}}} &= \overline{(x_i f_{x_i})} \cdot \overline{(\overline{x_i} f_{\overline{x_i}})} \\
&= (\overline{x_i} + \overline{f_{x_i}}) \cdot (x_i + \overline{f_{\overline{x_i}}}) \\
&= x_i \overline{f_{x_i}} + \overline{x_i} \overline{f_{\overline{x_i}}} + \overline{f_{x_i}}\ \overline{f_{\overline{x_i}}} \\
&= x_i(\overline{f_{x_i}} + \overline{f_{x_i}}\ \overline{f_{\overline{x_i}}}) + \overline{x_i}(\overline{f_{\overline{x_i}}} + \overline{f_{x_i}}\ \overline{f_{\overline{x_i}}}) \qquad \text{(absorption)} \\
&= x_i \overline{f_{x_i}} + \overline{x_i} \overline{f_{\overline{x_i}}}.
\end{aligned}
$$

Due to these equivalences, there are exactly four pairs of combinations which represent actually different functions, where the two combinations within a pair are functionally equivalent. Figure 7.8 illustrates this situation for the example of an OBDD node. Here, the edges carrying the complement attribute are labeled by an additional circle. Indeed, representatives of different pairs actually represent different functions.

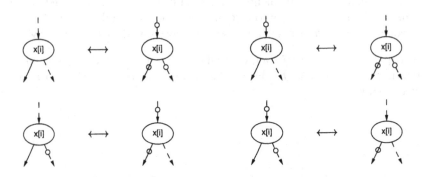

Figure 7.8. Pairs of equivalent representations

For this reason, one way to re-establish canonicity in the representation is the condition that the 1-edge of a node must never be complemented. Hence, in the equivalence pairs of Fig. 7.8 we always choose the left representative in each pair for uniquely representing the corresponding function. Whenever the OBDD package generates a new OBDD node, the edge attributes are chosen so that the condition is always satisfied.

Now it can be proven formally that this local strategy suffices to re-establish the global canonicity property. One has merely to take care that the edge to the root node of the OBDD may also be complemented. Otherwise, the constant 0-function cannot be represented as the complement of the 1-sink. The OBDD in Fig. 7.7 satisfies the presented restrictions to the positions of the complement bits. Furthermore, the reference to the root node of $\overline{f}$ is complemented. From now on, complemented 0-edges will always be drawn as dotted arrows as in Fig. 7.7.

The use of complemented edges provides several remarkable advantages:

- The compactness of the data structure is improved. The size of the OBDDs can at least theoretically be reduced up to 50%.
- The negation of a function can be performed in constant time.
- The application of Boolean operations can be significantly accelerated by exploiting such rules as, e.g., $f \cdot \overline{f} = 0$, $f + \overline{f} = 1$.

When using complemented edges, the complexity of the ITE algorithm can also be bounded quadratically if two of the arguments are complementary in each call. Since for all 16 binary operations, either this criterion or one of the criteria mentioned in the previous section can be applied, we can conclude the following theorem regarding the complexity of the ITE algorithm.

Theorem 7.5. *Let the switching functions f_1 and f_2 be represented by the OBDDs P_1, P_2 involving complemented edges with respect to the order π. For each binary operation the reduced OBDD P of $f = f_1 * f_2$ involving complemented edges can be determined in time $\mathcal{O}(size(P_1) \cdot size(P_2))$ by means of the ITE operator.* □

In practical work with OBDDs, it has turned out that in many cases the use of complemented edges only leads to a gain of about 10% in memory consumption. However, the possibility of complementing a function in constant time often leads to a gain of about a factor 2 in time consumption.

7.1.5 Standard Triples

Now, after introducing complemented edges, we return to the efficient realization of the ITE algorithm. It is possible that there are different function triples (f_1, f_2, f_3) and (g_1, g_2, g_3) with identical function values $\text{ITE}(f_1, f_2, f_3) = \text{ITE}(g_1, g_2, g_3)$. In order to achieve a high hit rate in the computed table, it is therefore advisable to transform each function triple to a normal form first. By means of these **standard triples** one avoids storing unnecessary triples in the computed table, and the recognition of equivalences helps to avoid unnecessary recomputations. Here, in particular, we shall exploit the advantages of complemented edges.

As an initial example, we consider some ITE calls which are all functionally equivalent to $f + g$:

$$\text{ITE}(f, f, g) = \text{ITE}(f, 1, g) = \text{ITE}(g, 1, f) = \text{ITE}(g, g, f).$$

We will now state a set of rules which transform a given triple into a standard form. The first series of transformation rules tries to replace functions by constants whenever possible:

$$\text{ITE}(f, g, g) \Rightarrow \text{ITE}(f, 1, g),$$
$$\text{ITE}(f, g, f) \Rightarrow \text{ITE}(f, g, 0),$$
$$\text{ITE}(f, g, \overline{f}) \Rightarrow \text{ITE}(f, g, 1),$$
$$\text{ITE}(f, \overline{f}, g) \Rightarrow \text{ITE}(f, 0, g).$$

In case of complemented edges, the test whether two of the occurring functions are equal or complementary to each other can be performed in constant time.

The terminal cases in the ITE recursion can now be extended by the following set:

$$\text{ITE}(f,1,0) = f,$$
$$\text{ITE}(f,0,1) = \overline{f},$$
$$\text{ITE}(1,f,g) = f,$$
$$\text{ITE}(0,f,g) = g,$$
$$\text{ITE}(f,g,g) = g.$$

The next series of transformation rules exploits the commutativity of the ITE operator if the second or the third argument are constant, or if they are complementary to each other. In these cases, one of the following identities holds:

$$\text{ITE}(f,1,g) = \text{ITE}(g,1,f),$$
$$\text{ITE}(f,g,0) = \text{ITE}(g,f,0),$$
$$\text{ITE}(f,g,1) = \text{ITE}(\overline{g},\overline{f},1),$$
$$\text{ITE}(f,0,g) = \text{ITE}(\overline{g},0,\overline{f}),$$
$$\text{ITE}(f,g,\overline{g}) = \text{ITE}(g,f,\overline{f}).$$

It is typically advisable to choose that form of a pair whose first argument depends on a variable which occurs earlier in the order. If the first argument of ITE is a single variable x_i which occurs before all variables of the other arguments, then the resulting node with label x_i can be immediately constructed. In this case, the other two arguments define the 1- and the 0-successor. Based on this observation we have another terminal case:

$$\text{ITE}(f,g,h) \;=\; (x_i,g,h) \quad \text{if } f = x_i$$
$$\text{and } x_i <_\pi topvar(g)$$
$$\text{and } x_i <_\pi topvar(h),$$

where $topvar(g)$ is the leading variable among those variables in the order π which are essential for g.

Another suitable series of standardization rules refers to the use of complemented edges. We have for example

$$\text{ITE}(f,g,h) = \text{ITE}(\overline{f},h,g)$$
$$= \overline{\text{ITE}(f,\overline{g},\overline{h})}$$
$$= \overline{\text{ITE}(\overline{f},\overline{h},\overline{g})}.$$

Each of the three functions f, g, h can be represented by a conventional or by a complemented edge. Among the four equivalent forms there is exactly one triple whose first two arguments are not referenced by a complemented edge. This triple is used as standard triple for looking up in the computed table, performing the ITE operation, and storing the result in the computed table. In case that one of the last two triples are chosen as standard triple, the computation yields the complement of the desired function. Hence, the resulting function must be complemented before returning it.

The stated rules which take care of complemented rules recognize equivalences due to DeMorgan's rules. For example, let us assume that the references to the functions f and g are not complemented. The chosen triple for the computation

$$f + g = \text{ITE}(f, 1, g)$$

is $\text{ITE}(f, 1, g)$, and the result is stored under this entry in the computed table. When a later request for

$$\overline{f} \cdot \overline{g} = \text{ITE}(\overline{f}, \overline{g}, 0) = \overline{\text{ITE}(f, 1, g)}$$

is carried out, then, once more, the standard triple $\text{ITE}(f, 1, g)$ is chosen, and the previously stored result can be recalled. According to the applied transformations, the result will be complemented before it is returned.

7.1.6 Memory Management

Large OBDDs typically emerge from combinations of smaller OBDDs by means of Boolean operations. Hence, in a typical application many small (intermediate) OBDDs are constructed which are only of temporary importance. In case of a symbolic simulation an OBDD is constructed for each gate of the circuit. Each of these OBDDs is only necessary until the OBDDs of all successor gates are computed. In order to achieve the naturally desirable deletion of those intermediate OBDDs which are no longer of use, one has to face the following problems:

- In general, an internal node in an OBDD has more than one predecessor. For space reasons the references to these predecessors cannot be stored in the node.
- The nodes in the OBDD are referenced from the unique table, from the computed table, and from other nodes.
- It does not seem reasonable to completely eliminate the intermediate nodes as soon as possible, as they could still be of great use for the caching mechanism of the computed table.

For these reasons, it seems to be a good strategy not to deallocate the memory cells of unused intermediate nodes immediately. Instead, it is better to wait until the necessary management overhead for restructuring compares well to the gain in storage. Such a strategy is called **garbage collection**.

In the framework of an OBDD package, garbage collection can be realized efficiently in the following way. For each node v we introduce a **reference**

counter which tells us how many nodes or external elements reference the node v. Whenever a new node is created whose 1-edge or 0-edge points to v, the counter is incremented. Whenever a function represented by v is no longer of use, the counter is decremented. If the counter reaches the value 0, then the reference counter of the sub-OBDDs in the two successors are decremented as well. Nodes whose reference counter is 0 are called **dead**. When the number of dead nodes is sufficiently large, garbage collection is activated. First the entries of dead nodes in the unique table and in the computed table are deleted. Subsequently, the memory cells used by the dead nodes are freed.

Even after the introduction of reference counters, the node v does not yet know which nodes are its predecessor nodes. However, at each moment it is known whether or not v is used any longer for the representation of other functions. Collecting dead nodes without immediately deallocating them offers the possibility of reviving dead nodes in case of need.

The costs of garbage collection amortize themselves when dealing with a relatively high number of dead nodes. Moreover, at the time of garbage collection the employed tables can be restructured quite easily. For example, the sizes of tables can be adapted dynamically to the actual requirements.

We conclude the discussion of memory management by considering the memory consumption of a single OBDD node again. At the beginning of this chapter, in Fig. 7.1, we described the basically necessary structure of a node. In the course of the chapter, we added some components for improving efficiency. In Section 7.1.2, the component **Next** was described that points to the successor in the collision list of the unique table.

The reference counter **Refcount** serves for organizing the garbage collection. Typically, it suffices to reserve between one and two bytes for this counter. In case of an overflow the node goes into a saturation state and will not be freed until the end of the application.

Furthermore many OBDD-based algorithms use some bits, e.g., for recording which part of the OBDD has already been visited. We designate a component **Mark** for this purpose. The extended data structure of a node is shown in Fig. 7.9. For a 32-bit architecture, the memory consumption per node is 4 memory words, i.e., 16 byte.

In case of using complemented edges, each edge is additionally supplied with an attribute bit. This single bit may be provided by means of a simple trick. If, as in the illustrated data structure, each node consists of exactly 4 memory words, then each node begins at an address whose last two bits are zero. Hence, one can set aside a bit for storing the complement attribute from the memory words **High** or **Low**.

In addition to the memory requirements of each individual OBDD node, we have to take into account the memory requirements of the tables whose sizes are dynamically adapted to the number of represented OBDD nodes. We

Component	Size
Index	2 byte
High	1 word
Low	1 word
Next	1 word
Refcount	ca. 2 byte
Mark	

Figure 7.9. Core of the record for representing an OBDD node

assume that the unique table and the computed table have the same size. A favorable strategy turns out to be to choose the size of the unique table by roughly a factor four smaller than the current number of nodes. In this case, each collision list contains four elements on average, and there is a good relation between memory consumption and access time. Each entry in the unique table is a pointer to a memory cell. In case of a 32-bit architecture, the amortized costs of this table are about 1 byte per node. Each entry in the computed table of the ITE operator consists of 4 memory words. Hence, in the case of a 32-bit architecture, the amortized costs of the computed table are about 4 byte per OBDD node.

Altogether, the memory consumption per node in the OBDD is about 21 byte. Given this estimate, an OBDD of 1 million nodes requires about 21 MB of memory.

7.2 Some Popular OBDD Packages

Within the last few years, numerous OBDD packages have been developed which provide interfaces for the manipulation of switching functions. Some of these packages have been created at academic institutions, and some in industrial development centers. Of course, the commercial packages are not, or only very restrictively, open to the public. However, as the development of OBDD technology has been driven strongly by universities, the publicly available academic packages provide a good insight into the state of the art. This holds even more, as some of the packages developed at academic institutions have been used in commercial CAD systems themselves.

7.2.1 The OBDD Package of Brace, Rudell, and Bryant

In the historical development, the first efficient implementation of the OBDD data structure in a program system called OBDD package was created by K. Brace at Carnegie Mellon University, in cooperation with R. Rudell and R. Bryant. The OBDD package was written in the time from 1989 to 1990, and offered to the public in 1990.

The aim of the development was to move the frontiers of applicability of the verification software *Tranalyze* as far as possible. Tranalyze serves for verifying transistor circuits in MOS (Metal-Oxide-Semiconductor) technology on the switching level. On this abstraction level, the individual transistors are modeled as switches. In the framework of the verification task a series of equivalence tests among the gates in a logic network are required. By introducing the OBDD data structure in the context of *Tranalyze*, the frontiers of the manageable circuits could be extended substantially.

Many of the implementation techniques described in this chapter were initially developed and applied by Brace, Rudell, and Bryant. The suitability of their design decisions is particularly confirmed by the observation that many of the newer OBDD packages still use the same principles.

7.2.2 The OBDD Package of Long

The experiences gathered with the OBDD package of Brace, Rudell, and Bryant were transferred to a new, improved package some time later. This package was designed and implemented by D. Long, also at Carnegie Mellon University. It was mainly focused on model checking applications, which will be treated in more detail in Chapter 11.

The package has been publicly available since 1993. In particular, the package has been integrated into the SIS software for sequential synthesis which was developed at the University of California at Berkeley, and which has already been mentioned in Chapter 5. Later on, the package of Long was also used in projects inside AT&T.

An important new feature in the package of Long was the inclusion of techniques for dynamically constructing good variable orders, which we will investigate in the next two chapters.

7.2.3 The CUDD Package: Colorado University Decision Diagrams

The **CUDD package** (Colorado University Decision Diagrams) was developed by F. Somenzi and his working group at the University of Colorado at Boulder. The initial version was made public in April 1996. By carefully and ingeniously redesigning the algorithms of previous packages, Somenzi achieved a series of substantial performance improvements.

The outstanding property of the CUDD package is a large collection of algorithms for improving the variable order. These algorithms will be discussed in the next chapters. Moreover, the package can handle several of the OBDD variants presented in Chapter 12, namely multi-terminal BDDs and zero-suppressed BDDs.

```
#include <stdio.h>
#include <stdlib.h>
#include <cudd.h>

int main (int argc, char *argv[]) {
/* Input: An integer n */
/* Output: Computation of an OBDD of
     x[1] x[2] + x[3] x[4] + ... + x[2n-1] x[2n] */

  DdManager *bddm
  DdNode *f, *tmp1, *tmp2;
  int i;

  if (argc < 2) return 0;
  else n = atoi(argv[1]);

  /* initialize manager */
  bddm = Cudd_Init(0, 0, CUDD_UNIQUE_SLOTS, CUDD_CACHE_SLOTS,
    CUDD_MAX_CACHE_SIZE);

  /* Initialize f to the zero function */
  f = Cudd_ReadLogicalZero(bddm);
  Cudd_Ref(f);

  for (i = 1; i <= n; i++) {
    tmp1 = Cudd_bddAnd(bddm,
      Cudd_bddIthVar(bddm, 2*i-1),
      Cudd_bddIthVar(bddm, 2*i));
    Cudd_Ref(tmp1);
    tmp2 = Cudd_bddOr(bddm, f, tmp1);
    Cudd_Ref(tmp2);
    Cudd_RecursiveDeref(bddm, f);
    f = tmp2;
  }

  printf("Size of the OBDD: %d \n", Cudd_DagSize(f));

  return 1;
}
```

Figure 7.10. Example of using CUDD

The CUDD package is used as the primary OBDD package in the VIS verification system, which will be discussed in Chapter 11.

Figure 7.10 shows how the program library in the programming language C can be used. The BDD manager *DdManager* contains all relevant data structures for the global organization of a shared OBDD, such as a unique table. In the CUDD package, for efficiency reasons, the user is responsible himself for incrementing the reference counter of newly constructed nodes. In this way, one can avoid the referencing and dereferencing process for intermediate functions that exist only very briefly. If a function is no longer of use, a command for recursive dereferencing is called.

7.3 References

The key ideas for the efficient implementation of OBDDs mostly go back to Brace, Rudell, and Bryant [BRB90]. Complemented edges were proposed by Karplus [Kar88] and by Madre and Billon [MB88]. The three presented OBDD packages are publicly available. The OBDD package of Long originated from the work [Lon93]. The CUDD package, the latest of these three packages, can be found at the Internet address given in reference [Som96b].

8. Influence of the Variable Order on the Complexity of OBDDs

Alle Räder stehen still,
wenn dein starker Arm es will.
[All wheels stand still
if your strong arm so will.]
Georg Herwegh (1817–1875)

In this chapter we analyze the influence of the variable order on the complexity of OBDDs. The following two theorems, which immediately follow from the Theorems and Corollaries 6.9 to 6.12, are applied several times.

Theorem 8.1. *Let S_j be the set of subfunctions of f which we obtain by fixing $x_1, \ldots, x_{j-1}$ to constants, and which depend essentially on x_j. The reduced OBDD of f with respect to the variable order $x_1, \ldots, x_n$ has exactly $|S_j|$ nodes labeled by x_j.* □

Theorem 8.2. *Let $(i_1, \ldots, i_n)$ be a permutation of the set $\{1, \ldots, n\}$, and let P be the reduced OBDD of f with respect to the variable order $x_{i_1}, \ldots, x_{i_n}$. Further, let S_{i_j} be the set of subfunctions of f which we obtain by fixing $x_{i_1}, \ldots, x_{i_{j-1}}$ to constants, and which depend essentially on x_{i_j}. Then the reduced OBDD P of f has exactly $|S_{i_j}|$ nodes labeled by x_{i_j}.* □

8.1 Connection Between Variable Order and OBDD Size

The size of an OBDD and hence the complexity of its manipulation depends on the underlying variable order – a dependency which can be quite strong. We want to consider some extreme examples.

The OBDD size of the function

$$f(x_1, \ldots, x_{2n}) = x_1 x_2 + x_3 x_4 + \ldots + x_{2n-1} x_{2n}$$

behaves very sensitively with regard to changes in the chosen variable order. With respect to the variable order $x_1, x_2, \ldots, x_{2n-1}, x_{2n}$ the reduced OBDD consists of exactly $2n + 2$ nodes. Hence, the growth is linear in the number

n of variables. The OBDD for the case $n = 3$ is depicted in Fig. 8.1 (a). The reason for the very compact OBDD follows from the fact that for each $k \in \{1, \dots, n-1\}$, after reading the assignments to the variables $x_1, \dots, x_{2k}$ there are only two possibilities:

1. Due to the assignments to $x_1, \dots, x_{2k}$ it is already known that the function value is 1.

2. The first possibility does not apply. By means of this information one can determine the function value by using only the assignments to the remaining variables $x_{2k+1}, \dots, x_{2n}$.

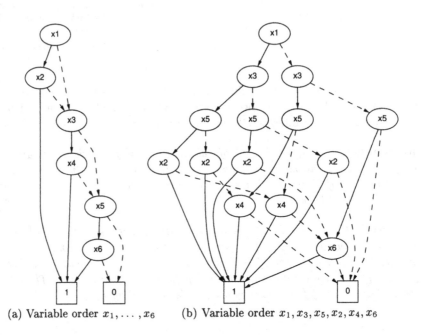

(a) Variable order $x_1, \dots, x_6$ (b) Variable order $x_1, x_3, x_5, x_2, x_4, x_6$

Figure 8.1. The function $x_1x_2 + x_3x_4 + x_5x_6$

In other words, due to Theorem 8.1 we have: For each k and arbitrary fixing of the variables $x_1, \dots, x_{k-1}$ to constants, there is always only one subfunction $f_k(x_k, \dots, x_n)$ which depends essentially on x_k. This subfunction is

$$f_k = \begin{cases} x_k x_{k+1} + x_{k+2} x_{k+3} + \dots + x_{2n-1} x_{2n} & \text{if } k \text{ odd,} \\ x_k + x_{k+1} x_{k+2} + \dots + x_{2n-1} x_{2n} & \text{if } k \text{ even.} \end{cases}$$

For the variable order $x_1, x_3, \dots, x_{2n-1}, x_2, x_4, \dots, x_{2n}$, the situation looks completely different. Figure 8.1 shows the corresponding OBDD for $n = 3$. We analyze the function by means of Theorems 8.1 and 8.2. First let $k \leq n$.

There exist 2^{k-1} different constant vectors $(a_1, a_3, \ldots, a_{2k-3}) \in \{0,1\}^n$ that can be assigned to the $k-1$ variables $x_1, x_3, \ldots, x_{2k-3}$. These vectors lead to the subfunctions

$$f(x_1, \ldots, x_n)|_{x_1=a_1, x_3=a_3, \ldots, x_{2k-3}=a_{2k-3}}$$
$$= a_1 x_2 + a_3 x_4 + \ldots + a_{2k-3} x_{2k-2} + a_{2k-1} x_{2k}$$
$$+ x_{2k+1} x_{2k+2} + \ldots + x_{2n-1} x_{2n} .$$

From this expression the following facts can be deduced:

1. Each of the 2^{k-1} subfunctions $f(x_1, \ldots, x_n)|_{x_1=a_1, x_3=a_3, \ldots, x_{2k-3}=a_{2k-3}}$ depends essentially on x_{2k-1}, the k-th variable in the order.
2. All of these subfunctions are pairwise different.

Consequently, in case $k \leq n$ the reduced OBDD of f has exactly 2^{k-1} nodes which are labeled by x_k. Analogously, it can be verified that in case $k > n$ there are exactly 2^{2n-k} nodes which are labeled by x_k. The total number of nodes amounts to

$$2 \cdot 2 \sum_{k=1}^{n} 2^{k-1} + 2 = 2 \cdot (2^n - 1) + 2 = 2^{n+1}.$$

Therefore the reduced OBDD of f with respect to this variable order grows exponentially in n. Of course, merely to show the exponential growth, it would have been sufficient to show the exponential growth of the nodes labeled by a specific variable x_i.

The intuitive reason for this strong growth in size is as follows. In contrast to the situation for the order $x_1, x_2, \ldots, x_n$, after reading variables $x_1, x_3, \ldots, x_k$ for some odd k, one cannot yet deduce any information about the function value in specific cases. For each assignment to the first variables the function value of f is not yet determined. Both of the values 0 and 1 are still possible by assigning suitable values to the remaining variables. This fact implies that no edge in the reduced OBDD from one of the above variables directly leads to a sink. However, it is even worse that for each of the two assignments to the first variables $x_1, x_3, \ldots, x_k$ an assignment to the remaining variables can always be found such that the resulting function values differ.

Symmetric functions. In case of symmetric functions the function value only depends on the number of 1's in the input vector, but not on their position. This property implies that all variable orders for the OBDD are equivalent and that the OBDD size is therefore *independent* of the chosen variable order. This and other statements concerning symmetric functions can be formally proven from Theorem 8.2. For this, let $f(x_1, \ldots, x_n)$ be a symmetric function. For each k and an assignment of constants to the $k-1$ variables $x_{i_1}, \ldots, x_{i_{k-1}}$ it is only important how many of these variables are 1. Hence, the number of different subfunctions

$$f_{x_{i_1}=a_{i_1},\ldots,x_{i_{k-1}}=a_{i_{k-1}}}$$

is always bounded by k. As a consequence, for each variable x_i the reduced OBDD of a symmetric function in n variables contains at most linearly many nodes labeled by that variable. Altogether, the reduced OBDD of a symmetric function has at most quadratic size in n. Hence, the reduced OBDD size of symmetric functions is not only independent of the variable order but is also quite small. Exponential growth cannot occur.

Many practically relevant functions are symmetric, a fact which has already been mentioned in Section 3.3.5. Hence, the small memory consumption of the OBDDs of symmetric functions is particularly pleasant. The OBDDs of the binary AND, OR and EX-OR functions were already depicted in Example 6.5. To explain the structure of specific symmetric functions, we therefore prefer to consider threshold functions

$$T_k^n(x_1,\ldots,x_n) = \begin{cases} 1 & \text{if } x_1 + \ldots + x_n \geq k, \\ 0 & \text{otherwise.} \end{cases}$$

After reading some arbitrary i variables which contain at least k ones, the function value is already determined. Let v be a node reached after an

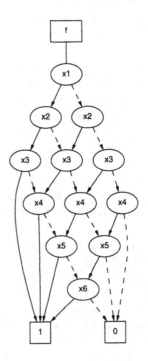

Figure 8.2. The majority function T_3^6

assignment to the first i variables that include at most $k-1$ ones. v contains a 1-edge that leads to the 1-sink. This further implies that for each i there are at most k nodes labeled by x_i. Figure 8.2 illustrates this fact in the example of the majority function T_3^6. In case of those two values of i, for which there are exactly 3 nodes labeled by x_i, these 3 nodes reflect the following information:

Left node: So far, exactly two 1's have been read.

Middle node: So far, exactly a single 1 has been read.

Right node: So far, no 1 has been read.

Adder. Let $f(a_{n-1}, b_{n-1}, \ldots, a_0, b_0) : \mathbb{B}^{2n} \to \mathbb{B}^n$ be a simplified adder function whose inputs are two n-bit numbers $a_{n-1} \ldots a_0$, $b_{n-1} \ldots b_0$, and which computes the last n bits of their binary sum $s_{n-1} \ldots s_0$, see Fig. 8.3. In particular, an overflow is not recognized. This adder can be seen as a switching function with n outputs.

a_{n-1}	a_{n-2}	$\ldots$	a_2	a_1	a_0
b_{n-1}	b_{n-2}	$\ldots$	b_2	b_1	b_0
s_{n-1}	s_{n-2}	$\ldots$	s_2	s_1	s_0

Figure 8.3. Simplified n-bit adder

Adders contain partial symmetries in the sense that for each i the variables a_i and b_i are completely equivalent. However, these partial symmetries cannot prevent the adder from depending very sensitively on the choice of the variable order. With respect to the variable order $a_{n-1}, b_{n-1}, \ldots, a_0, b_0$ the total size of the shared OBDD representing all n output bits is linear in n. In contrast to this, with respect to the variable order $a_{n-1}, \ldots, a_0, b_{n-1}, \ldots, b_0$ the size grows exponentially in n. Figure 8.4 elucidates this effect in case $n = 3$. In both of the depicted graphs complemented edges are used, as in this variant of OBDDs the structural properties of adders become more prominent.

8.2 Exponential Lower Bounds

We have seen that the dependency of the OBDD size on the underlying variable order can be very strong. It would be desirable if for all functions there were *at least* one variable order leading to a small OBDD. However, OBDDs share a fatal property with *all* representations of switching functions: the representations of nearly all functions require exponential space ! The proof

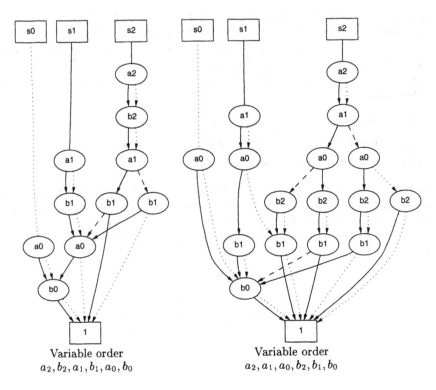

Variable order
$a_2, b_2, a_1, b_1, a_0, b_0$

Variable order
$a_2, a_1, a_0, b_2, b_1, b_0$

Figure 8.4. 3-bit adder function

of this fact is based on a counting argument which goes back to Shannon. The main idea is that the number of n-variable switching functions of 2^{2^n} is so huge that it is impossible under any circumstances to represent each of the functions in polynomial space. We will demonstrate the application of this counting argument for the example of OBDDs. However, the proof technique implies the generalization of the statement for *all* representations of switching functions.

Theorem 8.3. *We consider OBDDs with respect to their optimal order. Let $G(n)$ be the number of n-variable functions whose size is less than $2^n/2n$, and let $N(n) = 2^{2^n}$ be the number of all n-variable functions. Then the quotient $G(n)/N(n)$ converges to 0, as n tends to infinity:*

$$\frac{G(n)}{N(n)} \longrightarrow 0 \quad as \ n \longrightarrow \infty .$$

Proof. Let $K = \lfloor 2^n/2n \rfloor$. By c_i we denote the number of nodes in the OBDD which are labeled by x_i. Further let $c_{n+1} = K - (c_1 + c_2 + \ldots + c_n)$ denote the number of nodes which could theoretically be used in addition without violating the bound. As each number c_i is a nonnegative number, the equation $\sum_{i=1}^{n+1} c_i = K$ has exactly

$$\binom{n+K}{K}$$

solutions.

We assume that all nodes are enumerated successively in the way that the ordering property of the variable order is preserved: if x_i occurs before x_j in the order, then the number of each node labeled by x_i is smaller than the number of each node labeled by x_j. The two sinks are provided with the largest numbers.

Given this property of order preservation we can deduce that for each node the numbers assigned to its two sons are greater than its own number. As a consequence, there are at most

$$
\begin{aligned}
&((K-1)+2)^2 \cdot ((K-2)+2)^2 \cdot ((K-3)+2)^2 \cdot \ldots \cdot 2^2 \\
&= (K+1)^2 \cdot K^2 \cdot (K-1)^2 \cdot \ldots \cdot 2^2 \\
&= ((K+1)!)^2
\end{aligned}
$$

possible ways to place the edges. As there are $n!$ different variable orders, there exist at most

$$n!\binom{n+K}{K}((K+1)!)^2 = (K+1)(K+1)!(K+n)!$$

functions whose optimal OBDD consists of at most K nodes. The last expression is smaller than $(K+n)^{2K+n}/2^{2^n}$ which implies

$$
\begin{aligned}
\log_2(G(n)/N(n)) &\leq (2K+n)\log_2(K+n) - 2^n \\
&\leq ((2^n/n)+n)(n - \log_2 n + O(1)) - 2^n \\
&= n^2 + ((2^n/n)+n)(-\log_2 n + O(1)) \\
&= (2^n/n)\left(n^3/2^n + (1+n^2/2^n)(-\log_2 n + O(1))\right) \\
&\longrightarrow -\infty \quad \text{as } n \longrightarrow +\infty.
\end{aligned}
$$

Hence, the quotient $G(n)/N(n)$ converges to 0, as n tends to infinity. $\square$

The proof of the theorem is based on a pure counting argument and yields no information on what *concrete* examples of exponential OBDD sizes look like. Indeed, it is not at all easy to prove lower bounds for the resource requirements of concrete switching functions. For example, in the model of Boolean circuits the following situation can be found. Here, too, we have the theorem that nearly all functions can only be represented by Boolean circuits of exponential size. The question of what a concrete function with this property looks like is tightly connected with important consequences for fundamental problems in theoretical computer science. For this reason, many efforts have been made world wide to construct such a function. However, the result is sobering: nobody has succeeded yet in stating a concrete function with necessarily exponential circuit size. In fact, the situation is even worse: nobody has succeeded either in stating a concrete function with, say, at least quadratic size requirements in the circuit model. The "world record" concerning lower bounds for the size of circuits is the almost frighteningly small number of $3n$. This lower bound was proven by N. Blum in 1984 for an artificial function specifically constructed for this purpose.

Fortunately, in case of OBDDs the situation is not as hopeless as in the case of Boolean circuits. With respect to the proof of lower bounds the model of OBDDs can be handled in a significantly easier way. There are explicit constructions of switching functions whose OBDD representations have necessarily exponential resource consumption. These constructions do not require artificially and artistically created functions, but they can also be given in terms of practically relevant switching functions.

After the introduction of OBDDs by Bryant in 1986 and their successive incorporation into large CAD systems, nobody succeeded in constructing a good variable order for the binary multiplication function. At first, it was not clear whether fundamental reasons were the deciding factor, or merely the fact that the available optimization algorithms were not developed enough. It took several years until the answer was found, by proving that multiplication requires exponential OBDDs with respect to any variable order. This result from the year 1991 is also due to Bryant.

Definition 8.4. *By the term* **multiplication of n-ary binary numbers** *we denote the switching function*

$$F = F(x_{n-1}, \ldots, x_0, y_{n-1}, \ldots, y_0) : \mathbb{B}^{2n} \to \mathbb{B}^{2n},$$

whose inputs are two n-bit numbers $x = x_{n-1} \ldots x_0$ and $y = y_{n-1} \ldots y_0$, and which computes their binary product $z = z_{2n-1} \ldots z_0 = x \cdot y$.

The proof of the lower bound is based on methods of communication complexity. Let $f \in \mathbb{B}_n$ be a switching function over the set of variables $X = \{x_1, \ldots, x_n\}$, and let Y be a subset of X. By the term **subfunction of f on Y** we denote a subfunction which results from fixing all variables in $\overline{Y}$ to constants, where $\overline{Y}$ is the complement of Y.

Theorem 8.5. *Let $f \in \mathbb{B}_n$ be a switching function over the set of variables $X = \{x_1, \ldots, x_n\}$ with the following property: for all $Y \subset X$ of a fixed size m there are at least k different subfunctions on Y. Then each OBDD of f contains at least k nodes.*

Proof. We consider an OBDD of f with respect to an arbitrary variable order π. Denote the first $n - m$ variables in π by $\overline{Z}$ and the last m variables in π

by Z. Any two different assignments on $\overline{Z}$ which define different subfunctions on Z must lead to different functions in the OBDD. As the precondition of the theorem holds for any $Y \subset X$ of size m, it holds in particular for the chosen set Z. Hence, there are at least k different subfunctions on Z which lead to at least k different nodes in the OBDD. □

Now we apply this statement to families of functions $f_n \in \mathbb{B}_n$, $n \geq 1$, such as functions which compute one particular output bit of the n-ary multiplication function. In order to show that a family of functions f_n, $n \geq 1$, has necessarily exponential OBDDs, it suffices to prove that for a size $m = m(n)$ there exists an exponential number of subfunctions (e.g., 2^{cn} for some $c > 0$). To prove this large number of subfunctions one can proceed as follows. Show that for each subset $\overline{Y}$ of the variables that consists of $n - m$ elements there exists an exponential number of assignments to $\overline{Y}$ with the following property: for any two of these assignments there exists an assignment to the variables on Y that leads to different function values.

A useful notion in the formal treatment of this situation can be established by means of so-called **fooling sets**.

Definition 8.6. Let $f \in \mathbb{B}_n$ be a switching function over the set of variables $X = \{x_1, \ldots, x_n\}$. For $Y \subset X$ and $x, x' \in \{0,1\}$ the terms x_Y and $x_{\overline{Y}}$ denote the value of x on the variables in Y and $\overline{Y}$, respectively (analogous for x'). A set $F \subset \mathbb{B}^n$ is called **fooling set** of f with respect to Y if for all $x \neq x' \in F$ the following conditions are satisfied:

1. $f(x) = f(x') = 1$,
2. $f(x_Y x'_{\overline{Y}}) = 0$ or $f(x'_Y x_{\overline{Y}}) = 0$.

Theorem 8.7. Let $f \in \mathbb{B}_n$ be a switching function over the set of variables $X = \{x_1, \ldots, x_n\}$. If for all subsets $Y \subset X$ of a fixed size m the function f has a fooling set of size k, then each OBDD of f contains at least k nodes.

Proof. We consider an OBDD of f with respect to an arbitrary variable order π. Denote the first m variables in π by Y and the last $n - m$ variables in π by $\overline{Y}$. For two assignments $x \neq x'$ in the fooling set, the assignments x_Y and x'_Y on the first m variables cannot lead to the same node, as we have

1. $f(x) = f(x') = 1$,
2. $f(x_Y x'_{\overline{Y}}) = 0$ or $f(x'_Y x_{\overline{Y}}) = 0$,

and therefore

$$f(x_Y x_{\overline{Y}}) \neq f(x_Y x'_{\overline{Y}}) \text{ or } f(x'_Y x'_{\overline{Y}}) \neq f(x'_Y x_{\overline{Y}}) .$$

□

According to Theorem 8.7 the existence of fooling sets which have *exponential* size immediately implies an exponential lower bound on the size of the corresponding OBDDs. In case of multiplication we will now show that already the computation of the middle bit z_n of the multiplication function in Definition 8.4 requires exponential OBDDs. In fact, the middle bit is the "hardest" bit in multiplication in the sense that its computation is most difficult.

Theorem 8.8. *With respect to any order the (reduced) OBDD of the middle bit in the multiplication function grows exponentially in n.*

Proof. Let $f = f(x_{n-1}, \ldots, x_0, y_{n-1}, \ldots, y_0)$ be the middle bit z_n of the multiplication function in Definition 8.4. We show that with respect to each subset $S \subset \{x_0, \ldots, x_{n-1}\}$ of size $n/2$ the function f has a fooling set of size $2^n/8$. The elements of this fooling set will only differ in the assignments to the variables x_i. The variables y_i are fixed in a way that the multiplication is reduced to computing the sum of two integers. One of these integers corresponds to a subsequence of variables with low index, i.e., of variables in $\{x_0, \ldots, x_{n/2-1}\}$, and the other integer corresponds to a subsequence of variables with high index, i.e., of variables in $\{x_{n/2}, \ldots, x_{n-1}\}$. Then the outgoing carry bit of this addition exactly coincides with the function f.

We choose the two subsequences in a way that for each index i of the subsequences the following two conditions are satisfied:

1. For each i, the i-th bit of one of the two subsequences belongs to S, and the i-th bit of the other subsequence belongs to $\overline{S}$.

2. There is a $k \in \{1, \ldots, n\}$ such that the two partners of all chosen bit pairs have the same distance k.

In order to guarantee the two conditions, we define

$$S_L = S \cap \{x_0, \ldots, x_{n/2-1}\} \text{ and } S_H = S \cap \{x_{n/2}, \ldots, x_{n-1}\}$$

for the elements of S with low indices (S_L) and high indices (S_H). Analogously, the restrictions of $\overline{S}$ on the two domains are defined:

$$\overline{S}_L = \overline{S} \cap \{x_0, \ldots, x_{n/2-1}\} \text{ and } \overline{S}_H = \overline{S} \cap \{x_{n/2}, \ldots, x_{n-1}\}.$$

By using elementary analysis it can be easily shown that for any integers $0 \le a, b \le n/2$ with $a + b = n/2$,

$$a \cdot \left(\frac{n}{2} - b\right) + b \cdot \left(\frac{n}{2} - a\right) \ge \frac{n^2}{8}.$$

By setting $a = |S_L|$, $b = |S_H|$ this implies in our case that

$$|S_L \times \overline{S}_H| + |\overline{S}_L \times S_H| \ge \frac{n^2}{8}.$$

Hence, there are at least $n^2/8$ pairs which satisfy the first condition. We partition this set of pairs (x_i, x_j) according to the value $i - j$ (which satisfies $1 \le |i - j| < n$) in n parts. Subsequently, we choose the largest of these sets. This one consists of at least $n/8$ elements. Altogether we have found the two required subsequences which satisfy the two above mentioned properties. Figure 8.5 illustrates this connection with regard to the school method for multiplication.

$$
\begin{array}{ccccccccc|cccccccccl}
 & 0 & 0 & x_i & 1 & x_j & 1 & 1 & x_k & 0 & x_p & 0 & x_q & 0 & 0 & x_r & 0 & 0 & 0 & = x \\
\times & 0 & 0 & 0 & 0 & 0 & 0 & 0 & 0 & 0 & 1 & 0 & 0 & 0 & 0 & 0 & 0 & 1 & 0 & = y \\
\hline
 & 0 & 0 & x_i & 1 & x_j & 1 & 1 & x_k & 0 & x_p & 0 & x_q & 0 & 0 & x_r & 0 & 0 & 0 & \\
\cdots & x_k & 0 & x_p & 0 & x_q & 0 & 0 & x_r & 0 & 0 & 0 & & & & & & & & \\
\end{array}
$$

$$\uparrow$$
$$f(x, y)$$

Figure 8.5. There is a k (here 7) such that in each chosen pair of variables (here (x_i, x_p), (x_j, x_q), (x_k, x_r)) the variables have distance exactly k

Exactly two bits of $y_0, \ldots, y_{n-1}$ are set to 1. This is done in a way that in the pictorial representation, the elements of one of the chosen subsequences will be exactly below the elements of the other one. In Fig. 8.5, the variables of each pair (x_i, x_p), (x_j, x_q), (x_k, x_r) between the horizontal lines are in exactly the same horizontal position. The bits of $x_0, \ldots, x_{n-1}$ which have not been included in the subsequences are set to 1 if they occur between two variables of a sequence, and they are set to 0 otherwise. This implies that carries from one position to the next within the chosen sequence are propagated. For the specifically chosen assignment the middle bit of the multiplication exactly coincides with the carry bit in the addition of the two $n/8$-bit integers which are defined by the two subsequences.

Now we define a fooling set F for this addition with respect to S which automatically implies a fooling set for the whole multiplication, too. The addition can be imagined in the following way. One integer which is determined by the assignment on $\overline{S}$ is added to another one which is determined by the assignment on S.

To simplify the notation, we denote the variables of the subsequence which originate from S by $a = a_{n/8-1}, \ldots, a_0$, and the variables of the subsequence which results from $\overline{S}$ by $b = b_{n/8-1}, \ldots, b_0$. We define the fooling set by

$$
F = \left\{ (a_0, \ldots, a_{n/8-1}, b_0, \ldots, b_{n/8-1}) : \sum_{i=0}^{n/8-1} a_i 2^i + \sum_{i=0}^{n/8-1} b_i 2^i = 2^{n/8} \right\}.
$$

As for each $a \in \{0, \ldots, 2^{n/8-1}\}$ there is exactly one integer b such that a and b sum up to $2^{n/8}$, F is of cardinality $2^{n/8}$. Hence, the first condition in the definition of fooling sets is obviously satisfied. For the second condition it suffices to observe that in case $(a, b) \neq (a', b') \in F$ one of the two pairs (a, b') or (a', b) no longer sums up to $2^{n/8}$.   □

We would like to demonstrate the techniques for proving lower bounds by means of a second example. For this reason, we consider the *hidden weighted bit function*, which can be used to describe an indirect memory access.

Definition 8.9. *For* $x = (x_1, \ldots, x_n) \in \mathbb{B}^n$ *let* $wt(x)$ *denote the number of 1's in* x*. The **hidden weighted bit function** $HWB : \mathbb{B}^n \to \mathbb{B}$ *is defined as follows:*

$$HWB(x_1, \ldots, x_n) = \begin{cases} x_{wt(x)} & \text{if } wt(x) > 0, \\ 0 & \text{otherwise.} \end{cases}$$

For the representation of the hidden weighted bit function in terms of OBDDs the following statement is known.

Theorem 8.10. *With respect to any order the (reduced) OBDD of the hidden weighted bit function grows exponentially in the input length* n*.*

Proof. W.l.o.g. we can assume that n can be divided by 10. Otherwise, only some small technical considerations are necessary in addition. We show that for each subset $S = \{x_{j_1}, \ldots, x_{j_{|S|}}\}$ of the variables with $|S| = 0.6n$ an exponential number of assignments to the variables in S exists which lead to different subfunctions.

Let X_H and X_L be two sets of cardinality $0.4n$ which consists of variables with high and low indices, respectively.

$$X_H = \{x_{0.5n+1}, \ldots, x_{0.9n}\}, \tag{8.1}$$
$$X_L = \{x_{0.1n+1}, \ldots, x_{0.5n}\}. \tag{8.2}$$

The equation $|X_H| + |X_L| = 0.8n$ implies $|S \cap (X_H \cup X_L)| \geq 0.4n$ and further

$$|S \cap X_H| \geq 0.2n \quad \text{or} \quad |S \cap X_L| \geq 0.2n.$$

Hence, we can determine a set W of cardinality $0.2n$ which satisfies

$$W \subset S \cap X_H \quad \text{or} \quad W \subset S \cap X_L.$$

Case 1: $W \subset X_H$. Let the set F be defined by

$$F = \{x = (x_{j_1}, \ldots, x_{j_{|S|}}) \in \mathbb{B}^{|S|} : |\{x_i \in W : x_i = 1\}| = 0.1n \\ \text{and } \forall x_i \in S \setminus W : x_i = 1\}.$$

Each vector $x \in F$ defines an assignment to the variables in S which contains exactly $0.5n$ ones.

Case 2: $W \subset X_L$. Now let the set F be defined by

$$F = \{x = (x_{j_1}, \ldots, x_{j_{|S|}}) \in \mathbb{B}^{|S|} : |\{x_i \in W : x_i = 0\}| = 0.1n$$
$$\text{and } \forall x_i \in S \setminus W : x_i = 0\}.$$

Each vector $x \in F$ defines an assignment to the variables in S which contains exactly $0.1n$ ones.

In both cases Stirling's formula $n! \approx \sqrt{2\pi n}(n/e)^n$ implies the estimation

$$|F| = \binom{n/5}{n/10} = \Omega\left(\frac{2^{n/5}}{\sqrt{n}}\right) = \Omega\left(2^{(1/5-\epsilon)n}\right) = \Omega(1.14^n).$$

Now we show that each two assignments from the set F lead to different subfunctions. For this reason, let $x'_S \neq x_S \in F$ be assignments to the variables in S which differ in the i-th bit for some $i \in \{1, \ldots, n\}$, $x_i \neq x'_i$.

Case 1: $W \subset X_H$. Equation (8.1) implies $0.5n < i \leq 0.9n$. For an assignment $x_{\overline{S}}$ of the variables in $\overline{S}$, in which exactly $i - 0.5n$ bits are 1, we have $wt(x_S x_{\overline{S}}) = wt(x'_S x_{\overline{S}}) = i$, and hence

$$HWB(x_S x_{\overline{S}}) = x_i \neq x'_i = HWB(x'_S x_{\overline{S}}). \tag{8.3}$$

Case 2: $W \subset X_L$. Now $0.1n < i \leq 0.5n$. For an assignment $x_{\overline{S}}$ of the variables in $\overline{S}$, in which exactly $i - 0.1n$ bits are 1, we have $wt(x_S x_{\overline{S}}) = wt(x'_S x_{\overline{S}}) = i$ and hence also statement (8.3). $\qquad\square$

8.3 OBDDs with Different Variable Orders

From Chapter 6 we know that in case of two OBDDs with the *same* order, binary operations and the equivalence test can be performed efficiently. In the previous sections, we have therefore investigated problems under the precondition that at each moment the occurring OBDDs have the *same* fixed variable order.

Of course, one may also think of working with OBDDs with respect to different orders. For example, let P_1 be an OBDD with the variable order π_1 and P_2 be an OBDD with the variable order π_2. How efficiently can the basic operations still be performed ?

In this section, we first show that the equivalence test can also be performed in polynomial time for OBDDs with different variable orders. Note, however, that the resulting algorithm is significantly more costly than in case of identical orders. When investigating binary operations, the situation is quite different. We show that performing binary operations for OBDDs with different variable orders is an **NP**-hard problem. Exactly this property is the reason why in many cases it is required that all occurring OBDDs have a single common order.

```
Equivalence(P₁, P₂) {
/* Input: Read-once branching programs P₁, OBDD P₂
            with respect to the variable order π₂. */
/* Output: "Yes", if P₁ and P₂ represent the same function.
            "No", otherwise. */
    Initialize L to {(P₁, P₂)};
    For all nodes v in P₁ {
        visited[v] = FALSE;
    }
    Do {
        Let v be a node in P₁ with visited[v] = FALSE
            whose predecessors v' all satisfy the property visited[v'] = TRUE;
        Let v be labeled by a variable xᵢ, and
            let P₁ the subgraph in P rooted in v;
        Let {(P, Q₁), ... , (P, Qₖ)} be the list of pairs of graphs in L,
            in which P occurs;
        /* As Q₁, ..., Qₖ are subgraphs of an OBDDs with variable order π₂,
            their equivalence can be checked in polynomial time */
        If ¬(f_{Q₁} = ... = f_{Qₖ}) {
            Return( "No" );
        }
        If (P only consists of a sink) {
            /* As P trivially is an OBDD with variable order π₂, the
                equivalence of P and Q₁ can be checked in polynomial time */
            If ¬(f_P ≠ f_{Q₁}) {
                Return("No" );
            }
        } Else {
            Let P' and P" be the subgraphs in P₁ which are rooted
                in the 1- and 0-successor of P, respectively;
            Add the pairs (P', (Q₁)_{xᵢ=1}) and (P", (Q₁)_{xᵢ=0}) to L;
            Remove the pairs (P, Q₁), ..., (P, Qₖ) from L;
            visited[v] = TRUE;
        }
    } While (there exists a v with visited[v] = FALSE);
    Return ( "Yes" );
}
```

Figure 8.6. Equivalence test of OBDDs with different orders

Theorem 8.11. *Let P_1 be a read-once branching program and P_2 be an OBDD. Equivalence of P_1 and P_2 can be decided in polynomial time.*

Proof. Figure 8.6 contains an algorithm which decides the equivalence of P_1 and P_2. We show

1. The algorithm works correctly.
2. The algorithm runs in polynomial time.

The algorithm for the equivalence test is based on a simple principle: Obviously, for two switching functions $f, g \in \mathbb{B}_n$ and a variable x_i the following equation holds:

$$f = g \iff f_{x_i} = g_{x_i} \text{ and } f_{\overline{x_i}} = g_{\overline{x_i}}. \qquad (8.4)$$

The algorithms keeps a list L of branching programs with the following property:

$$f_{P_1} = f_{P_2} \iff \forall (P, Q) \in L \quad f_P = f_Q, \qquad (8.5)$$

where f_P and f_Q are the functions represented by P and Q, respectively.

At the beginning, L is initialized to $\{(P_1, P_2)\}$. In the processing of the algorithm L is only modified by the following kinds of operations:

1. A pair in L is replaced by the pairs of read-once branching programs of the cofactors according to condition (8.4).
2. For two pairs $(P, Q_1) \in L$ and $(P, Q_2) \in L$,

$$f_{Q_1} \neq f_{Q_2} \Longrightarrow f_{P_1} \neq f_{P_2},$$
$$f_{Q_1} = f_{Q_2} \Longrightarrow \text{one of two pairs can be removed from } L.$$

Hence, statement (8.5) is an invariant of the Do-loop. To prove the correctness of the algorithm, it therefore suffices to observe that the invariant holds at the beginning of the algorithm, and that at the end of the processing one of the following conditions is satisfied:

1. For all nodes v in P_1 it holds that $visited[v] = \text{TRUE}$, the list L is empty, and the answer is "Yes".
2. There is a node v in P_1 with $visited[v] = \text{FALSE}$ as well as a pair (P, Q) with $f_P \neq f_Q$, and the answer is "No".

Regarding the running time, for each node in P_1 the Do-loop is executed at most once. Each of these executions requires at most a number of $size(P_2)$ many equivalence tests of OBDDs with the same order. Each of these tests can be performed in polynomial time. Hence, the running time of the algorithm in Fig. 8.6 is polynomial. □

As an OBDD is a special case of a read-once branching program, we can immediately deduce the following corollary.

Corollary 8.12. *Equivalence of two OBDDs with different variable orders can be decided in polynomial time.* □

Now we show that the application of binary operations to OBDDs with different variable orders is an **NP**-hard problem.

Problem 8.13. The problem **COMMON_PATH$_{\pi_1\text{-OBDD},\pi_2\text{-OBDD}}$** is defined as follows:

Input: An OBDD P_1 with order π_1 and an OBDD P_2 with order π_2.

Output: "Yes", if there is an assignment to the input variables such that P_1 and P_2 simultaneously compute the value 1. "No", otherwise.

Lemma 8.14. *The problem* **COMMON_PATH$_{\pi_1\text{-OBDD},\pi_2\text{-OBDD}}$** *is* **NP**-*complete.*

Proof. As each of the two OBDDs can be evaluated in polynomial time, the problem is in **NP**.

To show that the problem is **NP**-hard, it suffices to analyze an earlier proof again, this time in more detail. In Theorem 4.32 it has been proven that satisfiability of ordered read-2 branching programs of width 3 is an **NP**-complete problem.

The proof idea was as follows: For each instance C of the **NP**-complete problem 3-SAT/3-OCCURRENCES we have constructed two branching programs $P'(C)$ and $P''(C)$ with the following property: C can be satisfied if and only if there is a variable assignment to $P'(C)$ and $P''(C)$ such that $P'(C)$ and $P''(C)$ simultaneously compute the value 1. From the construction rules for $P'(C)$ and $P''(C)$ it can be immediately recognized that both programs even satisfy the OBDD property. As a consequence, the problem COMMON_PATH$_{\pi_1\text{-OBDD},\pi_2\text{-OBDD}}$ is **NP**-hard.   □

Problem 8.15. The problem **·-SYN$_{\pi_1\text{-OBDD},\pi_2\text{-OBDD}}$** is defined as follows:

Input: An OBDD P_1 with order π_1 and an OBDD P_2 with order π_2.

Output: An OBDD P (with arbitrary variable order) of the function $f = f_1 \cdot f_2$, where f_1 and f_2 denote the functions which are computed by P_1 and P_2, respectively.

Theorem 8.16. *The problem* **·-SYN$_{\pi_1\text{-}OBDD,\pi_2\text{-}OBDD}$** *is* **NP**-*hard.*

We proceed analogously to the proof of Theorem 4.38.

Proof. The problem COMMON_PATH$_{\pi_1\text{-OBDD},\pi_2\text{-OBDD}}$ can be solved by applying an algorithm for ·-SYN$_{\pi_1\text{-OBDD},\pi_2\text{-OBDD}}$ to the input OBDDs, and then deciding SAT$_{\text{OBDD}}$ on the resulting OBDD. Due to Lemma 8.14 COMMON_PATH$_{\pi_1\text{-OBDD},\pi_2\text{-OBDD}}$ is an **NP**-complete problem, whereas SAT$_{\text{OBDD}}$ can be decided in polynomial time. Consequently, the investigated problem ·-SYN$_{\pi_1\text{-OBDD},\pi_2\text{-OBDD}}$ is **NP**-hard.   □

As complementation is quite easy in the context of OBDDs, one can immediately obtain the following corollary.

Corollary 8.17. *The two problems $+\text{-}SYN_{\pi_1\text{-}OBDD,\pi_2\text{-}OBDD}$ as well as $\oplus\text{-}SYN_{\pi_1\text{-}OBDD,\pi_2\text{-}OBDD}$ are **NP**-hard.* □

After the analysis of the two fundamental operations we would like to go briefly into the problem of transforming an OBDD P_1 with an order π_1 into an OBDD P_2 with an order π_2. From Section 8.1 it is known that in this process the size of an OBDD can increase by an exponential factor. Hence, there cannot be an algorithm – not even a nondeterministic (!) one – which guarantees to perform this transformation in polynomial time. Hence, the best that one can hope for, is an algorithm whose running time is bounded by a polynomial in $size(P_1)$ and $size(P_2)$. The next theorem, which we state without proof, implies that such an algorithm exists. Here, one speaks of a **global rebuilding** of the OBDD.

Theorem 8.18. *Let P_1 be an OBDD with variable order π_1, and let π_2 be another variable order. There is an algorithm which transforms P_1 into a reduced OBDD P_2 with variable order π_2, and whose running time is bounded by $\mathcal{O}(size(P_2)^2 \cdot size(P_1) \cdot \log(size(P_1)))$.*

8.4 Complexity of Minimization

As the size of an OBDD very strongly depends on the chosen variable order, algorithms for constructing good orders are of great practical importance. But the hope that there is an *efficient* algorithm which finds the best among all possible orders cannot be satisfied. In this section it is shown that the test whether the minimal OBDD size of a function (given also in terms of an OBDD) is smaller than a given number s is **NP**-complete. Hence, it is an **NP**-hard problem to construct the optimal order to a given OBDD, and efficient algorithms cannot be expected.

In the historical development the exact clarification of the complexity theoretical questions was a complicated adventure. First, in the original paper of Bryant in 1986 it was stated without proof that the construction of an optimal order is an **NP**-hard problem. This plausible statement was accepted as fact for some time. However, as the role of the variable order in various applications became increasingly important, it was investigated in more detail, and it turned out that the complexity of the ordering questions had not been solved at all. Indeed, for a long time this aspect eluded clarification.

Finally, in 1993 the Japanese trio S. Tani, K. Hamaguchi, and S. Yajima succeeded partially by proving the relevant complexity statement for the somewhat weaker case of shared OBDDs:

Problem 8.19. The problem **OPTIMAL SHARED OBDD** is defined as follows:

Input: A shared OBDD P and a positive integer s.

Output: "Yes", if the functions which are represented by P can be represented by shared OBDD P' (with respect to an arbitrary variable order) with at most s nodes. "No", otherwise.

Theorem 8.20. *The problem OPTIMAL SHARED OBDD is* **NP***-complete.*

By means of the techniques they developed the authors did not succeed in transferring the hardness result to the more specific class of OBDDs (i.e., shared OBDDs with exactly one root), a class that may be easier to handle in the complexity theoretical sense. However, in 1996, B. Bollig and I. Wegener managed to close this gap. They proved the complexity statements also for the class of OBDDs.

Problem 8.21. The problem **OPTIMAL OBDD** is defined as follows:

Input: An OBDD P and a positive number s.

Output: "Yes", if the function being represented by P can be represented by an OBDD P' (with respect to an arbitrary variable order) with at most s nodes. "No", otherwise.

Theorem 8.22. *The problem OPTIMAL OBDD is* **NP***-complete.*

The proofs of the two theorems are quite long and rather technical, a fact that reflects the complicated nature of the seemingly very elementary structures. However, the basic ideas of both proofs give fundamental insights into the tools for theoretical analysis of OBDDs. Hence, we would like to provide a sketch of the proof for the case of shared OBDDs.

For the proof of Theorem 8.20 we first prove that the problem OPTIMAL SHARED OBDD is in the class **NP**. Of course, this statement implies that the more specific problem OPTIMAL OBDD is in the class **NP**, too.

Theorem 8.23. *The problem OPTIMAL SHARED OBDD is in* **NP**.

Proof. The OBDD P' with at most s nodes can be guessed. By Corollary 8.12 the equivalence of P' and the input OBDD P can be checked in polynomial time.   □

Theorem 8.24. *The problem OPTIMAL SHARED OBDD is* **NP***-hard.*

Proof. For the proof we reduce the well-known **NP**-complete problem OPTIMAL LINEAR ARRANGEMENT (OLA) to the problem OPTIMAL SHARED OBDD.

OPTIMAL LINEAR ARRANGEMENT:

Input: An undirected graph $G = (V = \{1, \ldots, n\}, E)$ and a positive integer K.

Output: "Yes", if there is a permutation ψ on $\{1, \ldots, n\}$ such that

$$\sum_{(u,v) \in E} |\psi(u) - \psi(v)| \leq K.$$

"No", otherwise. The expression $|\psi(u) - \psi(v)|$ is called the *costs of the edge* (u, v), and the sum $\sum_{(u,v) \in E} |\psi(u) - \psi(v)|$ is called the *costs of the graph G*.

The basic idea of the reduction is to reflect the costs of an edge in the graph by the size of an OBDD. The OBDDs which are constructed in this way constitute elementary components within the reduction. The relevant shared OBDDs are assembled from these elementary components.

Definition 8.25. *A (u, v)-**phage function** in the variables $x_1, \ldots, x_n$ is a switching function of the form*

$$f(x_1, \ldots, x_n) = (x_u \oplus x_v) \prod_{k \notin \{u,v\}} x_k. \tag{8.6}$$

Figure 8.7 shows a phage function with respect to several variable orders, where edges to the 0-sink are omitted. If $\Delta(u, v)$ denotes the distance between the two variables x_u and x_v in the order, then the phage function (8.6) has exactly $n + \Delta(u, v)$ internal nodes.

Hence, for each linear arrangement ψ of a graph G the following holds. If the edge(u, v) has costs of C, then the OBDD of the (u, v)-phage function (8.6) with respect to the variable order $x_{\psi^{-1}(1)}, \ldots, x_{\psi^{-1}(n)}$ has exactly $C + n$ internal nodes. In this way, the desired correspondence is established. Figure 8.8 shows the costs of an edge for two different linearly arranged graphs. The two phage functions which correspond to these two orders are exactly those in Fig. 8.7.

Note that in general, the graph contains not just a single edge but many edges. If all corresponding phage functions depend on the same set of variables then subgraphs can be shared. As a consequence, the size of the shared OBDD no longer reflects the costs of the arrangement.

Things can be put right by considering a separate variable set for each edge and hence for each phage function: for the k-th edge the phage function is defined on the set of variables $x_{1,k}, \ldots, x_{n,k}$. Hence, altogether we have $n \cdot |E|$ variables. If for each j the variables $x_{j,1}, x_{j,2}, \ldots, x_{j,|E|}$ are consecutive successors in the variable order, then the size of the resulting shared OBDD corresponds exactly with the costs of the arrangement. If for each j the shared OBDD satisfies this condition to the variable order, we call it *well-ordered*.

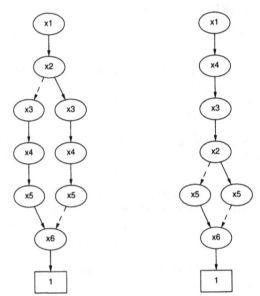

Variable order $x_1, \ldots, x_6$ Variable order $x_1, x_4, x_3, x_2, x_5, x_6$

Figure 8.7. The phage function $(x_2 \oplus x_5)\, x_1 x_3 x_4 x_6$

Costs of the edge: 3   Costs of the edge: 1

Figure 8.8. An edge with respect to two different linear arrangements

To establish a well-ordered OBDD, we add a modified phage function for each node j, called the *penalty function*,

$$h_j = (x_{j,1} \oplus x_{j,2} \oplus \ldots \oplus x_{j,|E|}) \prod_{i,l \neq j} \overline{x_{i,l}}.$$

The penalty function causes that the variables $x_{j,1}, \ldots, x_{j,|E|}$ are kept together in the order. In the presented modeling there are still some small difficulties which occur because of possible sharing of subgraphs in the lowest OBDD level. These difficulties can be removed by doubling the set of involved variables for each edge.

Definition 8.26. *An **OLA decision diagram** of a graph (V, E) given in the problem OPTIMAL LINEAR ARRANGEMENT is a shared OBDD which represents the subsequently defined $2|E| + n$ functions f_k, g_k, $1 \le k \le |E|$, and h_j, $1 \le j \le n$. If the k-th edge of E is labeled by (u, v), then*

$$f_k = (x_{u,k} \oplus x_{v,k}) \prod_{i \notin \{u,v\}} x_{i,k},$$

$$g_k = (y_{u,k} \oplus y_{v,k}) \prod_{i \notin \{u,v\}} y_{i,k}.$$

For each node j the function h_j is defined by

$$h_j = (x_{j,1} \oplus x_{j,2} \oplus \ldots \oplus x_{j,|E|} \oplus y_{j,1} \oplus y_{j,2} \oplus \ldots \oplus y_{j,|E|}) \prod_{i,l \neq j} \overline{x_{i,l}} \cdot \overline{y_{i,l}}.$$

Due to possible sharing of subgraphs in the lowest OBDD level, the size of a well-ordered OLA decision diagram is not determined uniquely. Instead of this, there are two possible types which differ in their size by exactly one node. Exactly for solving this problem we have doubled the set of variables. The following two statements can be proven formally with some efforts.

Lemma 8.27. *For the OLA decision diagram of a linearly arranged graph $G = (V, E)$ with costs K the following holds: if the OLA decision diagram is well-ordered then its size with respect to the types 1 and 2 amounts to*

$$k_1(K) = k_2(K) + 1 = 2 \left(K + 1 - |V| + |E||V|(|V| + 2) - |E| \sum_{i=1}^{|V|-2} i \right),$$

where k_i refers to type i. □

Lemma 8.28. *Each non-well-ordered OLA decision diagram can be transformed into a functionally equivalent well-ordered OLA decision diagram with fewer nodes.* □

Here, we omit the proofs of the two lemmas and look instead at the overall conclusion of the reduction. For this purpose we consider an instance of OPTIMAL LINEAR ARRANGEMENT. This instance consists of a graph and a positive integer s. We transform the graph into a well-ordered OLA decision diagram and compute the number $k_1(s)$ according to Lemma 8.27. Then we use the OLA decision graph P and the value $k_1(s)$ as input to the problem OPTIMAL SHARED OBDD. If there is no equivalent shared OBDD to P which has at most $k_1(s)$ nodes, then there is no linear arrangement with costs at most s.

We now assume that an OLA decision diagram with at most $k_1(s)$ nodes exist which represents the same function as the well-ordered decision diagram. If the OLA decision diagram is well-ordered, then there exists a linear arrangement of the original graph G with costs s. (As $k_2(s) < k_1(s) < k_2(s+1)$ for each positive integer s, OLA decision diagrams of type 2 do not cause problems either). If the OLA decision graph is not well-ordered, then Lemma 8.28

guarantees the existence of a well-ordered OLA decision diagram with at most $k_1(s) - 1$ nodes. Hence, there exists a linear arrangement ψ with costs at most s. As the whole construction can be performed in polynomial time, the problem OPTIMAL LINEAR ARRANGEMENT has been reduced to OPTIMAL SHARED OBDD in polynomial time. This implies the claim.

$\square$

8.5 References

The asymptotic statements about switching functions go back to Shannon [Sha49]. The mentioned paper concerning a linear lower bound for circuits is by Blum [Blu84]. The proofs of the exponential lower bounds of multiplication as well as for the hidden weighted bit function can be found in [Bry91]. Our treatment follows the presentation of Ponzio [Pon95].

The polynomial algorithm for performing the equivalence test of OBDDs with different orders was designed by Fortune, Hopcroft, and Schmidt [FHS78]. Gergov and Meinel proved that performing binary operations on OBDDs with different orders is **NP**-hard [GM94b]. Efficient algorithms for global rebuilding were designed independently by Meinel and Slobodová [MS94], by Savický and Wegener [SW97], and by Tani and Imai [TI94].

The two mentioned papers concerning the complexity of the minimization problems are [THY93, BW96].

9. Optimizing the Variable Order

Ordnung führt zu allen Tugenden.
Was aber führt zur Ordnung ?
[Order leads to all virtues.
But what leads to order ?]
Georg Christoph Lichtenberg (1742–1799)

Before representing and manipulating switching functions in terms of OBDDs, an order on the set of variables has to be fixed. In the previous chapter, we have seen that the construction of the optimal variable order is a very critical venture – as it is related to exploding running times. A good order can lead to a very compact representation and hence to small running times, whereas a bad representation may exceed the physically existing memory and hence causes the whole computation to abort. Even in the cases where bad variable orders do not cause a memory overflow, they lead to unacceptably large running times.

9.1 Heuristics for Constructing Good Variable Orders

Already in the first research papers about OBDDs the choice of suitable variable orders was investigated, and some general rules for manually choosing a good order were presented. However, the popularity of OBDDs strongly increased exactly at the moment when powerful heuristic methods became available to deduce a priori information for determining a good order.

The methods for heuristically constructing good orders can be classified by the available input information. On the one hand, the switching functions under consideration may be given in form of net lists, Boolean formulas, disjunctive normal forms, or other representation types, which are not sufficient for *manipulating* the functions. On the other hand, additional information may be provided and can be exploited, e.g., information concerning the semantic meaning of certain input bits of a function. A third, quite frequent situation is that the relevant functions are already provided in the form of OBDDs.

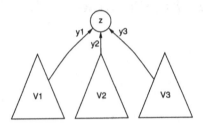

Figure 9.1. Motivation for the fan-in heuristic

Of course, the heuristic algorithms for constructing good variable orders depend on the type of input information. The main application in which these constructions were studied is the symbolic simulation of a combinational circuit, see Section 5.2. Starting from a given circuit representation the represented function is to be converted into an OBDD. The idea of the heuristics is to deduce information concerning suitable positions of the variables in the order from the topological structure of the circuit. In the following, two of these heuristics, containing typical ideas, are described in more detail: the **fan-in heuristic** of Malik, Wang, Brayton, and Sangiovanni-Vincentelli, and the **weight heuristic** of Minato, Ishiura, and Yajima.

9.1.1 The Fan-In Heuristic

Definition 9.1. *Let v be a node in a combinational circuit C (which we consider as a directed acyclic graph). The **transitive predecessor cone** of v denotes all nodes w with the property that there is a path of positive length to v. The **reflexive-transitive predecessor cone** also contains the node v itself.*

As motivation for the fan-in heuristic we consider the circuit structure in Fig. 9.1. We assume that the reflexive-transitive predecessor cone V_i of the predecessor nodes y_i of a node z are pairwise disjoint. In order to compute the function value in z for a given input, we need to know the function values of all predecessors of z. Let y be one of these predecessors. If the values of all variables in the reflexive-transitive predecessor cone of y are known, then the function value in y can be deduced. For determining the function value in z the values of these variables are no longer necessary, because

1. the function value in y is already known, and
2. the relevant variables do not occur in the other predecessor cones.

If the variables which appear within the same predecessor cone occur consecutively in the order, then few subfunctions exist, and the OBDD remains small. For this reason, the fan-in heuristic proposes to keep those variables which appear within the same predecessor cone consecutively in the order.

Although in general, the predecessor cones are not disjoint, the realization of this idea leads to a provably useful heuristic. All gates of the given circuit are traversed by means of depth first search starting in the output node. If there are several output nodes, they are first combined to a single pseudo output function by a gate (e.g., the parity of all output nodes). The traversal in depth first search guarantees that variables within the same predecessor cone occur in the order as close to each other as possible.

The question which predecessor cone of a node should be traversed first is answered by the following idea. It seems to be advisable to read those variables in the OBDD first that appear far away from the output nodes in the circuit. Intuitively, these variables can influence the behavior of the circuit quite strongly. Reading these variables early in the OBDD allows representation of different behaviors, being induced by different values of the variables, in different sub-OBDDs. In particular, this avoids the need to remember the current "state" of *all* occurring behaviors in each level of the OBDD *simultaneously* (which would be quite costly). The formal realization of this idea is achieved by defining the depth of a gate in a circuit:

Definition 9.2. *Let C be a combinational circuit (which we consider again as a directed, acyclic graph with some specific nodes: the output gates). For each gate v of C the* **TFI-depth** *(transitive fan-in) is defined as follows:*

$$
TFI\text{-}Depth(v) = \begin{cases} 0 & \text{if } v \text{ is an output node,} \\ 1 + \max\{TFI\text{-}Depth(w) : \\ \quad w \text{ is a successor of } v\} & \text{otherwise.} \end{cases}
$$

Pseudo code for the realization of the fan-in heuristic is given in Fig. 9.2. First, the function *FanIn* calls the recursive function *FanInOrder* in which depth first search is performed. If the current node has not been visited yet, then, first, the predecessor nodes are considered recursively, and, next, the node is appended at the end of the list of visited nodes.

Altogether, that order on the variables is chosen which is consistent with the computed list of nodes *nodelist*. The earlier an input node labeled by a variable is reached, the earlier this variable appears in the order.

9.1.2 The Weight Heuristic

The **dynamic weight assignment heuristic**, for short the **weight heuristic**, also tries to exploit the topology of a circuit C in order to deduce information about a suitable variable order for the function computed by C.

```
FanIn(C) {
/* Input: A combinational circuit C with input variables
               {x₁,...,xₙ} and exactly one output z. */
/* Output: DFS traversal of all nodes in C. */
    nodelist = ∅;
    FanInOrder(z, nodelist);
}


FanInOrder(node y, list of nodes nodelist) {
    If (y ∉ nodelist) {
           Let {y₁,...,yₖ} be the set of predecessor nodes of y;
           For i = 1,...,k {
               tᵢ = TFI-Depth(yᵢ);
           }
           Let L = (l₁,...,lₖ) be the list of nodes yᵢ
               sorted by decreasing keys tᵢ;
           For i = 1,...,k {
               FanInOrder(lᵢ, nodelist);
           }
           Append(y, nodelist);
    }
}
```

Figure 9.2. Fan-in heuristic

In contrast to the fan-in heuristic, which is based on local decisions, the approach of the weight heuristic is more *globally* oriented.

Starting from the output, each node is assigned a weight which measures the influence of the node on the circuit. The variable with the largest weight is put in the first position of the variable order under construction. Initially, the weight 1 is assigned to each output node.

The weight of a node is propagated to the input nodes according to the following two rules:

1. At each gate, the weight of the gate is divided equally and distributed to the predecessor gates.
2. If a gate has several successor gates, all weights which are assigned to the gate (due to the first rule) are accumulated.

Formally, these rules can be captured by means of the following definitions.

Definition 9.3. *Let C be a combinational circuit. For each gate v of C we define the **weight** of v as follows:*

$$
Weight(v) = \begin{cases} 1 & \text{if } v \text{ is an output node,} \\ \sum_{w \text{ successor of } v} Weight(w)/indegree(w) & \text{otherwise.} \end{cases}
$$

After computing the weights of all nodes the input variable with the largest value is determined. Intuitively, this variable has a large influence on the represented function, and it is put at the beginning of the order. Subsequently, we remove the part of the circuit which is influenced by the chosen variable. The weight assignment is computed again in order to determine the second variable in the order. By repeated application of weight assignment and removal the complete order is successively computed. An example of the application of the weight heuristic is shown in Fig. 9.3. In the first assignment, the variable d has the largest weight. After deleting the corresponding circuit part the former weight of the variable d is redistributed. Intuitively, this redistribution causes that variables which are somehow related to the variable d obtain a super-proportional weight increase. In the depicted example the variable c has the largest weight in the second assignment, and hence, it is put at the second position in the order.

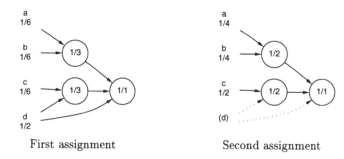

First assignment   Second assignment

Figure 9.3. The weight heuristic

9.2 Dynamic Reordering

The ordering heuristics described in the previous sections lead to good results for many circuits, but they have also some drawbacks:

1. The methods are quite problem specific. For example, the presented heuristics exploit the topology of the input circuit. However, if the relevant switching functions do not origin from a circuit representation but from another input source, there may be much less structural information.

2. The methods work *heuristically*, and in general, they are not able to find the optimal order. Although in many cases the resulting quality of heuristically constructed orders is sufficient, there may also be extreme cases where the heuristics fail.

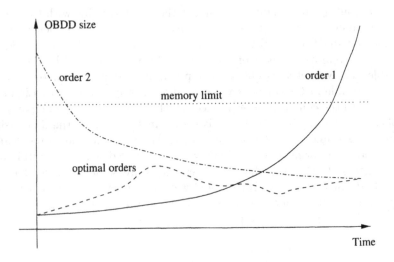

Figure 9.4. The problem of static orders

3. The heuristics produce a static order which remains unchanged during the whole course of the application. Many applications, such as the analysis of finite state machines, are not amenable to such a static approach. In this case, the optimal order at the beginning of the application can be completely different from the optimal order at the end. If one works with one and the same order for the whole time, one is often doomed to failure. Figure 9.4 illustrates this effect. Here, order 1 is the optimal order of the OBDDs at the *beginning* of the application, and order 2 is the optimal order of the OBDDs at the *end* of the application. The OBDD sizes referring to the optimal order at each moment are shown by the third line – note that the optimal orders at different times may be different. In many cases, when working with a fixed order it is not possible to stay within the memory limits. In the case of such an overflow, the complete computation fails.

A solution of this dilemma is to improve the variable order *dynamically* in the course of processing the given manipulation task. Here, one speaks of *dynamic reordering*.

Like the heuristics, these methods can also be classified by the type of input information which can be used. Concerning the dynamic application, one is particularly interested in those methods which achieve suitable reordering by merely using the current OBDD representation. These algorithms try to determine an improved variable π' from a given OBDD with an initial order π. As according to the results of the previous chapter the computation of an optimal order is an **NP**-hard problem, one typically does not aim at computing an *optimal* order, but one is satisfied with sufficiently large improvements.

The use of a reordering algorithm can now be controlled as follows:

Explicit calls: The user controls the points in time at which a reordering step is to be performed. For example, this may be desirable before beginning to perform a complex operation.

Automatic calls: The reordering algorithm is called automatically whenever certain situations arise. Typically, a reordering step is called whenever the size of the shared OBDD representation has been doubled since the last reordering call.

The reordering aspect is particularly interesting, as it can run in the background without direct interaction to the application program. Typically, for the application program only the references to the represented functions are of interest, and not their internal representation. The dynamic adjustment of the variable order allows the internally performed OBDD representation to be hidden from the outside almost completely.

9.2.1 The Variable Swap

A central observation which forms the key idea in many dynamic reordering algorithms is that *two neighboring variables in the order can be swapped efficiently*. Of course, this statement is not valid in an unrestricted manner, but depends on the chosen implementation. For this reason, we refer in the following to the basic framework described in Chapter 7, upon which nearly all existing OBDD packages are based. For this framework we show how a swap of two neighboring variables in the order can be realized efficiently.

First, we assume that the variable x_i occurs immediately before the variable x_j in the order. The effect of this swap on each node labeled by x_i can be seen by applying Shannon's expansion with respect to x_i and x_j. If the function which is represented in a node with label x_i is denoted by f, then we have:

$$f = x_i x_j f_{11} + x_i \overline{x_j} f_{10} + \overline{x_i} x_j f_{01} + \overline{x_i}\ \overline{x_j} f_{00}.$$

By using commutativity we can order the terms so that that x_j occurs before x_i:

$$f = x_j x_i f_{11} + x_j \overline{x_i} f_{01} + \overline{x_j} x_i f_{10} + \overline{x_j}\ \overline{x_i} f_{00}.$$

In other words, the actual effect of the swap is the exchange of the two subfunctions f_{10} and f_{01} in the OBDD. Here, we have to take care that all unconcerned nodes in the graph are not affected by this exchange.

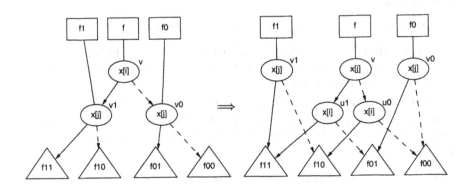

Figure 9.5. Swapping two neighboring variables

Figure 9.5 illustrates the swap of the two neighboring variables x_i and x_j within an arbitrary, possibly quite large OBDD. In the initial order the function f is represented by a node v labeled by x_i. First, we consider the case that the two sons v_1 and v_0 of v are labeled by x_j. The successor nodes of v_1 and v_0 represent the sub-OBDDs of the cofactors f_{11}, f_{10}, f_{01} and f_{00}. As f depends on the variable x_j, after the modification of the order this function has to be represented by a node with label x_j. The 1-successor of this node must possess references to the sub-OBDDs f_{11} and f_{01}, the 0-successor must have references to the sub-OBDDs f_{10} and f_{00}.

Note that the function f in Fig. 9.5 is represented by the same node v before and after the swap. Only the label and the outgoing edges of the node have been modified. This strategy guarantees that all existing references to f are not affected by the swap: neither the references which result from the upper levels in the OBDD, nor the references from outside the OBDD. As also the two cofactors f_1 and f_0 of f are represented by the original nodes v_1 and v_0 after the swap, *each* existing reference remains valid.

It is merely necessary to introduce the nodes u_1 and u_0 which represent the cofactors of f with respect to x_j. The figure seems to express that the size of an OBDD always increases during a variable swap. In case of a reduced representation, the equivalent status of x_i and x_j tell us that this cannot hold true. Indeed, there are even two reasons which reflect this general equivalence in the realization:

Nodes u_1, u_0: It is possible that the cofactors of f with respect to x_j are already represented in the original OBDD.

Nodes v_0, v_1: The preservation of the nodes v_1 and v_0 is only necessary if besides the original reference starting in the node v there is at least one other

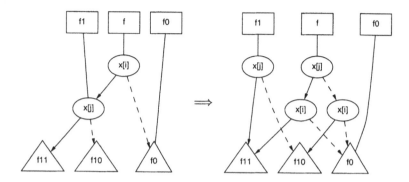

Figure 9.6. Special case of the variable swap where f_0 does not depend on x_j

reference. In a typical implementation these references cannot be efficiently determined, but the number of these references *can* be efficiently determined (see Section 7.1.6). If the reference counter of v_1 is zero after deleting the reference from the node v, then v_1 can be removed. The same holds true for v_0.

In the special cases where at least one of the two successor nodes of v is not labeled by x_j analogous constructions can be performed. For example, let the cofactor f_0 be independent of the variable x_j. By means of the construction in Fig. 9.6 the special case can be performed in such a way that all existing references remain valid.

Memory management. During performing a variable swap many dead nodes can arise. This suggests connecting the procedure directly with a garbage collection. That connection can be achieved as follows. Before starting an algorithm based on variable swaps a garbage collection is called, and the contents of the computed table is deleted such that all dead nodes are deleted. When swapping two neighboring variables x_i and x_j, only the reference counters of the nodes u_1 and u_0 may be decreased. If a reference counter reaches the value zero, then the node is removed immediately. In this way, it is guaranteed in a typical memory management framework that during dynamic reordering no dead nodes are carried along.

Complemented edges. In case of OBDDs with complemented edges the variable swap can be realized analogously. Here, it may happen at first that during the construction a 1-edge obtains the complement bit. But this can be corrected quickly and locally. We assume that not the subfunction f_{10} itself but instead its complement is represented. Hence, the edge to the sub-OBDD f_{10} carries the complement bit. Swapping the variables first leads to the graph in Fig. 9.7. The OBDD contains a complemented 1-edge which

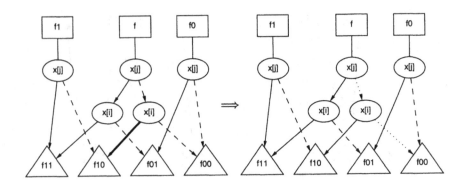

Figure 9.7. Variable swap in case of complemented edges

is drawn as a bold arrow. The figure also shows the transformation which serves to remove the complement bit from the 1-edge. Here, it is important that no unconcerned edge is affected by this transformation. At the end of the construction uniqueness of the representation has been re-established.

Time consumption. The efficiency of the variable swap substantially depends on the time needed for accessing the set of *all* nodes with label x_i. The time- and space-efficient framework in the form presented in Chapter 7 does not allow one to realize this access efficiently. However, a small modification can change this.

Let us recall that the access to the nodes is performed by means of a unique table. From each node with label x_i we have a very fast access to its sons, but not to all the other nodes with label x_i. However, by introducing a separate unique table for each variable x_i, this situation changes. Now, by using the collision lists we have fast access to the set of *all* nodes with label x_i. The required time for this access is

$$\mathcal{O}(\#lists + \#nodes),$$

where $\#lists$ is the number of collision lists in the unique table, and $\#nodes$ is the number of nodes in the table. Typically, the number of nodes is greater than the number of collision lists, so all nodes with label x_i can be visited in linear time.

As for each node with label x_i only constantly many operations are performed, the variable swap can be performed in linear time with regard to number of nodes with label x_i.

9.2.2 Exact Minimization

In most cases computing the optimal order of a given OBDD is too costly; nevertheless, exact methods are also worth considering. In particular, they are necessary

- to judge the optimization quality of heuristics or dynamic reordering algorithms,
- to work with the optimal order in particularly critical cases, or
- to be able to apply the exact method at least to parts of the OBDDs within a heuristic procedure.

The best technique so far for finding the optimal variable order was proposed by S. Friedman and K. Supowit, and is based on the principle of **dynamic programming**. When applying this programming paradigm the investigated problem is decomposed into several subproblems. The subproblems are solved, and their results are all stored in tables. Finally, the initial problem is solved by suitable look-ups in the constructed tables.

In the algorithm of Friedman and Supowit, the subfunctions of the OBDD constitute the mentioned subproblems. The procedure has running time $O(n^2 \cdot 3^n)$. This running time is exponential, but it is significantly better than the naive method of trying all $n!$ different orders successively.

In the context of dynamic reordering mechanisms, we would like to present a variant of the algorithm which has been proposed by N. Ishiura, H. Sawada, and S. Yajima. The three authors have taken up and realized the ideas of Friedman and Supowit. In contrast to the original method, the explicit construction of the tables is omitted, and instead of this, all relevant subfunctions are represented by OBDDs. The construction of currently not available subfunctions is carried out solely by using the variable swaps presented in Section 9.2.1.

Let $f \in \mathbb{B}_n$ be a switching function over the set of variables $\{x_1, \dots, x_n\}$. By OBDD(f, π) we denote the OBDD of f with respect to the order π, and by $cost_x(f, \pi)$ the number of nodes with label x in OBDD(f, π). Furthermore, for a subset $I \subset \{x_1, \dots, x_n\}$ the set $\Pi[I]$ consists of all orders π which satisfy

$$\{\pi[n - |I| + 1], \pi[n - |I| + 2], \dots, \pi[n]\} = I.$$

For the optimization algorithm the following observation is of central importance.

Lemma 9.4. Let $f = f(x_1, \dots, x_n) \in \mathbb{B}_n$, $I \subset \{x_1, \dots, x_n\}$ and $x \in I$. Then $cost_x(f, \pi)$ is invariant for all orders $\pi \in \Pi[I]$ with $\pi[n - |I| + 1] = x$.

OptimizeExact(P, π) {
/* Input: An OBDD P with respect to the variable order π */
/* Output: The OBDD size $OptCost$ with respect to the optimal
              variable order π' of the function represented by P */
    $MinCost_\emptyset = 0$;
    $\pi_\emptyset = \pi$;
    For $k = 1, \ldots, n$ {
         For each subset $I \subset \{1, \ldots, n\}$ of size k {
              Compute $MinCost_I$ and π_I by using
                   $MinCost_{I \setminus \{x\}}$ and $\pi_{I \setminus \{x\}}$ $(x \in I)$;
         }
    }
    $OptCost = MinCost_{\{1, \ldots, n\}}$;
}

Figure 9.8. Exact minimization

Proof. Theorem 8.1 implies: For each order $\pi \in \Pi[I]$ with $\pi[n - |I| + 1] = x$ the value $cost_x(f, \pi)$ is exactly the number of subfunctions which depend essentially on x, and which result from fixing all variables in $\{x_1, \ldots, x_n\} \setminus I$ to constants. Of course, this value is independent of the chosen order in which the variables are fixed.   □

The lemma says that the number of nodes with label x does not change when the order of the variables above (resp. below) of x is modified. By π_I we denote the order in $\Pi[I]$ which minimizes

$$\sum_{x \in I} cost_x(f, \pi),$$

and by $MinCost_I$ we denote the corresponding minimal value of this sum. The algorithm for exactly computing the optimal variable order can be algorithmically realized in the following way:

1. We assume inductively that for each subset of I which contains exactly $k - 1$ elements the optimal order π_I is known. π_I describes the optimal order on the set I if the variables in I are read in the bottom part of the OBDD.

2. By means of the orders for the subsets of size $k - 1$, the orders π_I for the subsets of I that contain exactly k elements can be computed. Figure 9.8 shows pseudo code for the algorithm. For the construction of π_I, each of the variables in I is successively put to the $(n - k + 1)$-th position in the order. Lemma 9.4 implies that for each of these variables it is optimal to construct the order in such a way that the variables in $I \setminus \{x\}$ are ordered according to $\pi_{I \setminus \{x\}}$. The corresponding costs can be determined by a simple accumulation.

```
ComputeMinCost {
/* Input: A subset I of the variables, already computed OBDD(f, π_{I\{v}}) */
/* Output: MinCost_I, the number of nodes in OBDD(f, π_I) */
    MinCost_I = ∞;
    For each variable x ∈ I {
        Reconstruct OBDD(f, < π_{I\{x}}, x >) by using OBDD(f, π_{I\{x}});
        NewCost_x = Cost_x(f, < π_{I\{x}}, x >) + MinCost_{I\{x}};
        If NewCost_x < MinCost_I {
            MinCost_I = NewCost_x;
            π_I = < π_{I\{x}}, x >;
            OBDD(f, π_I) = OBDD(f, < π_{I\{x}}, x >);
        }
    }
}
```

Figure 9.9. Computing the minimal costs

Figure 9.8 contains the main loop of the exact minimization method, and Figure 9.9 gives pseudo code for the computation of π_I. The process of appending a variable x to an existing part of an order π is abbreviated by the notation $< \pi, x >$.

Implementation by means of variable swaps. The algorithm in the presented form illustrates the gap between time complexity and space complexity, which can also be observed in many other algorithms. In a time-efficient implementation, *all* orders π_I of size $k - 1$ are stored until the computation of all orders π_I of size k has been completed. In this case the OBDD of each order of size k can be generated by successive variable swaps which merely move the currently chosen variable x to the $n - |I| + 1$-th position.

By contrast, in a more space-efficient implementation of this idea one would not store the OBDDs themselves, but only the orders. From this, the construction of an OBDD with respect to a specific variable order can still be performed by a sequence of variable swaps. However, in general many variables have to be moved to the right position, so the time consumption is substantially greater.

Lower bounds. By computing lower bounds for the size of OBDDs, in some situations the procedure can be interrupted in advance. A suitable lower bound is the following one. Let $MinCost_I$ and π_I be already computed for a subset I of the variables, and let c denote the number of nodes which are labeled by the variable $\pi[n - |I| + 1]$. These c nodes with label $\pi[n - |I| + 1]$ imply that there have to be at least $c - 1$ nodes above this level in the OBDD. Hence, a lower bound of the OBDD size, which can be established when starting from the (part of the) order π_I, is $MinCost_I + c - 1$. If this value is greater than the previously best computed total size this searching branch does not have to be pursued any further. The lower bound is particularly effective when the exact optimization algorithm is started from an already good order.

9.2.3 Window Permutations

Now we turn to algorithms which try to improve the order of a given OBDD without aiming at the optimum. The **window permutation algorithm** is based on the observation that during iterations of variable swaps one quite easily gets stuck in local minima. In order to prevent this, mechanisms are provided to exchange k neighboring variables simultaneously for a $k > 2$. Of course, the computation efforts increase strongly with increasing values of k. The algorithm successively traverses all levels $i \in \{1, \ldots, n - k + 1\}$ and systematically tries all $k!$ permutations of the variables $\pi[i], \ldots, \pi[i + k - 1]$. Subsequently, the permutation where the smallest OBDD was achieved is reconstructed, and an analogous step is performed for the next window of variables.

Here, the $k!$ permutations on k variables should be generated by the minimal possible number of $k! - 1$ swaps. For $k = 3$ and the variable notations a, b, c, the following sequence can be chosen:

$$abc \to bac \to bca \to cba \to cab \to acb.$$

For $k = 4$ and the variables a, b, c, d the following sequence achieves the desired effect:

$$abcd \to bacd \to badc \to abdc \to adbc \to adcb \to dacb \to dabc$$
$$\to dbac \to bdac \to bdca \to dbca \to dcba \to dcab \to cdab \to cdba$$
$$\to cbda \to bcda \to bcad \to cbad \to cabd \to cadb \to acdb \to acbd.$$

The method can be extended in a natural way to arbitrary values of k. For each positive integer k it is possible to generate the $k!$ different permutations of length k by using only $k! - 1$ swaps. After trying all permutations the best one is reconstructed. For this, only $k(k - 1)/2$ additional variable swaps are required.

In each case, the size of the resulting OBDD is never greater than the size of the initial OBDD. If the OBDD size actually decreases the algorithm can of course be iterated until no further improvement is achieved. Here, additional flags for the individual windows can be useful to avoid hopeless trials. The flag of a window is set after the optimal permutation within the window has been determined. The flag is unset again if for one of the preceding $k - 1$ windows a new permutation is determined. If during the course of the algorithm one hits upon a window where the flag is already set, this window does not have to be investigated again. If the flags of all windows are set, the algorithm will not be able to achieve any further improvements.

9.2.4 The Sifting Algorithm

In 1993, R. Rudell proposed the **sifting algorithm** for the dynamic minimization of OBDDs. The basic idea of this algorithm is to overcome the following weaknesses in the *window permutation algorithm*:

- Sometimes, many iterations are needed to move a variable to a position far away from its position in the initial order.
- Optimizing the order locally within the windows often implies that the optimization process gets stuck in a *local* minimum even after iterated application of the algorithm. In fact, this local minimum may still be far away from the global optimum.

The *sifting algorithm* of Rudell aims to eliminate these problems. The method is mainly based on the application of a subroutine which looks for the best position of a given variable without changing the positions of the other variables. This subroutine is successively applied to all variables $x_1, \dots, x_n$ in the order. In detail, for the current variable x_i the following two steps are performed:

1. The variable is moved through the whole order, and the minimum of the OBDD size in this process is recorded (see Fig. 9.10).
2. The variable is placed at the position in the order where the minimum in the first step was observed.

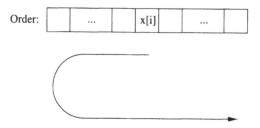

Figure 9.10. Idea of the sifting algorithm

During step 1 the size of the OBDD may increase. If in the course of moving a variable the size of the OBDD strongly increases, it becomes less and less probable that moving the variable further in this direction may produce a minimum. For this reason, the searching routine of the current variable is interrupted if the size increase of the OBDD exceeds a given factor *MaxGrowth*.

The pseudo code of a basic variant of the sifting algorithm is provided in Fig. 9.11. OBDD-size(P, π) denotes the size of the (shared) OBDD P with

---

```
Sifting(P_0, π_0) {
/* Input: An OBDD P_0 and a variable order π_0 */
/* Output: An OBDD P and a variable order π with
            OBDD-size(P, π) ≤ OBDD-size(P_0, π_0) */
    P = P_0; π = π_0;
    For all i ∈ {1, ... , n} {
            1.(a) /* Move variable x_i through the order */
                optsize = OBDD-size(P, π);
                optpos = curpos = startpos = π^{-1}[i];
                For j = startpos − 1, ... , 1 (decreasing) {
                    curpos = j;
                    swap_π(P, x_{π[j]}, x_{π[j+1]});
                    If OBDD-size(P, π) < optsize {
                        optsize = OBDD-size(P, π);
                        optpos = j;
                    }
                    Else If OBDD-size(P, π) > MaxGrowth * optsize {
                        Exit(Step 1.(a));
                    }
                }
            1.(b) For j = curpos + 1, ... , n {
                    curpos = j;
                    swap_π(P, x_{π[j−1]}, x_{π[j]});
                    If OBDD-size(P, π) < optsize {
                        optsize = OBDD-size(P, π);
                        optpos = j;
                    }
                    Else If OBDD-size(P, π) > MaxGrowth * optsize {
                        Exit(Step 1.(b));
                    }
                }
            2. /* Put variable x_i at the best position found in step 1 */
                If curpos > optpos {
                    For j = curpos − 1, ... , optpos (decreasing) {
                        swap_π(P, x_{π[j]}, x_{π[j+1]});
                    }
                }
                Else {
                    For j = curpos + 1, ... , optpos {
                        swap_π(P, x_{π[j−1]}, x_{π[j]});
                    }
                }
    }
}
```

Figure 9.11. Basic variant of the sifting algorithm

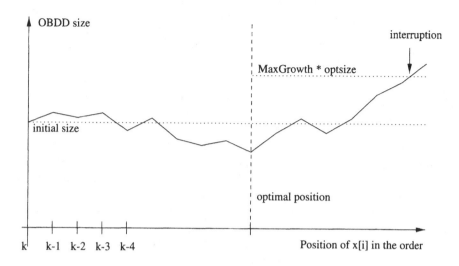

Figure 9.12. Example of a change in size when moving the variable x_i in the order. The initial position of x_i is k. During the searching process a new minimum is found. If the size increase exceeds the factor *MaxGrowth*, then the further movement of x_i in the current direction is interrupted

respect to the variable order π. For $1 \le i \le n$ the term $\pi[i]$ denotes the index of the variable at position i in the order, and $\pi^{-1}[j]$ denotes the position of the variable x_j in the order. The algorithm uses a subroutine $swap_\pi(P, x_i, x_j)$, which swaps the variables x_i and x_j in the order, and adjusts the OBDD P as well as the arrays $\pi[\,]$ and $\pi^{-1}[\,]$ accordingly.

The effectiveness of the sifting algorithm is based on the capability to move a variable quickly over a long distance in the order. It may occur quite often that the OBDD size first increases, and not until later drops below the initial value (see Fig. 9.12). In particular, this property allows it to escape from a local size minimum within the optimization space. The position at which the current variable is placed depends only on the determined minimum, and is independent of possibly traversed intermediate minima.

So far, the idea of sifting has turned out to be the best approach for constructing good variable orders in practical applications. For this reason, besides algorithmic refinements, which we shall discuss in Section 9.2.5, *efficient implementations* of sifting-based methods have been investigated in great detail. Here, it is not the aim to improve the general asymptotic running time behavior, but to improve the CPU times within practical applications. Consideration of the following ideas has turned out to lead to substantially better running times.

Factor MaxGrowth. An obvious implementation detail is how to make a suitable choice of the interruption factor *MaxGrowth*. In the original paper of Rudell a factor *MaxGrowth* = 2 was proposed. However, experimental studies have shown that the stricter factor *MaxGrowth* = 1.2 typically leads to a big gain in time without losing too much of the optimization quality.

Order of variable consideration. The sifting order, in which the variables are considered successively, already has a large influence on the optimization process. A useful heuristic consists of sorting the variables according to decreasing occurrence frequencies first. In other words, the first variable to be investigated is the one which occurs most frequently as label of a node, and therefore possesses the largest optimization potential.

Moving direction. In the presented basic variant the current variable is first moved to the top in the OBDD (i.e., decreasing in the order) and then to the bottom. As Fig. 9.13 shows, this method may not be suited for variables that appear right at the end of the order. For example, the second variable from the end has to be moved twice through the whole order before the optimal position of the variable is determined. Hence, as a typical heuristic a variable is first moved to the nearer end.

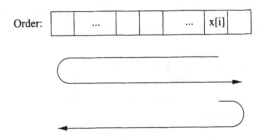

Figure 9.13. When "sifting" the second variable from the end in the order it is favorable to move the variable to the right end of the order first

Interaction matrix. The interaction matrix is a useful tool to guarantee high speed for the sifting algorithm. This Boolean $n \times n$-matrix, where n denotes the number of variables, is constructed before entering the main sifting loop. The entry (i, j) in the matrix contains a 1 if and only if among the functions $f_1, \dots, f_m$ represented in the shared OBDD there is a function f_i which depends essentially on both variables x_i and x_j. In other words, the entry (i, j) contains a 1 if and only if there is a root in the OBDD from which one can reach nodes with label x_i as well as nodes with label x_j. If two variables x_i and x_j do not interact in this sense, then in particular there is no edge which leads from a node with label x_i to a node with label x_j. Hence, all necessary adjustments in the realization of a variable swap of x_i and x_j are restricted to the description of the order. In the shared OBDD itself no modifications have to be performed at all. In case of two non-interacting variables, a variable swap can therefore be performed even in constant time, independently of the number of nodes labeled by x_i or x_j.

Lower bounds. By using lower bounds like those in the discussion of exact minimization (see Section 9.2.2), in some cases the searching step in the sifting algorithm can be interrupted in advance. When moving the variable x_i, we check some simple lower bounds at each traversed position. This tells us whether a further movement of x_i in the current direction can – at least in principle – still improve the minimum. If this is not the case, the movement of x_i in the current direction can be interrupted immediately.

9.2.5 Block Sifting and Symmetric Sifting

Of course, there are also problematic cases for the sifting algorithm. One critical aspect is that the absolute position of a variable is the main optimization criterion. In contrast to this, the relative positions within certain variable sets are only indirectly taken into account.

For illustrating the related difficulties, we consider a function in two variables a and b which have a strong attraction to each other in the following sense. Each good variable order of the function requires that the distance between a and b in the order is not too large. After the variable a is moved through the order within the sifting process, it is placed near b with high probability, as b "attracts" the variable a. If, later on, the variable b is moved through the order, it is placed near a for analogous reasons. Nevertheless, the position of the variable group $\{a, b\}$ may be far away from a really good position for this group. This effect of attraction is called the *rubber band effect*.

Variable groups in which the function is partially symmetric have this property of mutual attraction. Indeed, often the sifting algorithm presented above does not find the optimal position of such groups of symmetric variables.

Example 9.5. The interval function $I_{2,2}(x_3, \ldots, x_6)$ yields a 1 if exactly two of the four input bits are 1. Hence, the function

$$f(x_1, \ldots, x_6) = (x_1 \oplus x_2) \cdot I_{2,2}(x_3, \ldots, x_6)$$

is symmetric both in the variable pair $\{x_1, x_2\}$ and in the variable set $\{x_1, x_2, x_3, x_4\}$. The described rubber band effect can be well recognized if we start from the variable order $x_1, \ldots, x_6$ and observe what happens if the variable x_2 is moved to the bottom of the OBDD. Figure 9.14 shows both the initial situation and the situation where x_2 is at the fourth position in the order. Up to the level of the variable x_2 in the OBDD the value of the variable x_1 has to be kept available, so the size of the OBDD increases with the distance of the variables x_1 and x_2 in the order. $\diamond$

One solution of this problem is to unite mutually attracting variables into a group. Then, within the sifting algorithm, instead of single variables the

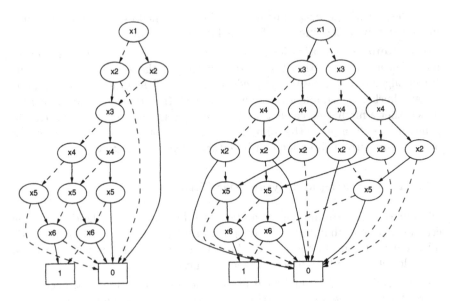

x_2 at the second position x_2 at the fourth position

Figure 9.14. The rubber band effect

whole group is moved through the order. For the definition of the groups we distinguish between the following two strategies:

- The groups are defined a priori by the user. Here, the user can employ his or her knowledge of the application.
- The groups are automatically constructed during the sifting process.

The first strategy is denoted as **block sifting**. If a function describes a system which consists of several modules, it may be advisable to unite the variables of each module a priori into a block. After this definition of the blocks, there are two different possible sifting-based reordering strategies:

1. The individual blocks are reordered by means of a sifting strategy.
2. The variables *within* each block are reordered by means of a sifting strategy.

The term **symmetric sifting** refers to a sifting variant introduced by S. Panda and F. Somenzi which realizes the second strategy. It is the aim of this method to unite symmetric variable groups during the sifting procedure automatically. Furthermore, variables whose behavior is similar to the behavior of a symmetric variable group are also united into a block. The basis of this unification during the sifting process is constituted by the following two statements. The first of these statements immediately follows from the definition of partial symmetry.

Lemma 9.6. *A switching function $f(x_1, \ldots, x_n)$ is partially symmetric in x_i and x_j if and only if:*

$$(f_{x_i})_{\overline{x_j}} = (f_{\overline{x_i}})_{x_j}.$$

$\square$

The second lemma now tells us how symmetry with respect to neighboring variables in the OBDD can be determined effectively.

Lemma 9.7. *Let P be the OBDD of a switching function $f(x_1, \ldots, x_n)$ with respect to the order π, and let x_i and x_j be neighbors in π with $x_i <_\pi x_j$. The function f is symmetric in x_i and x_j if and only if the following two statements hold true.*

1. *For all subfunctions g represented by nodes with label x_i we have:*

$$(g_{x_i})_{\overline{x_j}} = (g_{\overline{x_i}})_{x_j}.$$

2. *All edges which lead to a node with label x_j originate in a node with label x_i.*

Proof. If f is symmetric in x_i and x_j, then the condition of Lemma 9.6 holds true. This property can be transferred to all subfunctions of f, in particular to the nodes with label x_i. The proof of the second property is carried out by contradiction. We assume that there were an edge which leads to a node with label x_i, and which does not originate in a node with label x_i. Then there is a subfunction h of f with $h_{x_i} = h_{\overline{x_i}}$ which depends essentially on x_j, i.e.,

$$(h_{\overline{x_i}})_{x_j} \neq (h_{\overline{x_i}})_{\overline{x_j}}.$$

This implies

$$(h_{x_i})_{\overline{x_j}} = (h_{\overline{x_i}})_{\overline{x_j}} \neq (h_{\overline{x_i}})_{x_j},$$

in contradiction to the symmetry precondition.

The reverse direction can be shown analogously. $\square$

By using these criteria it can be efficiently checked during the sifting process whether two variables are symmetric.

In the implementation of symmetric sifting within the CUDD package, variables are also united if the conditions in Lemma 9.7 are violated for a small percentage of nodes and edges. Furthermore, so-called negative symmetries are considered, too. In contrast to "conventional" symmetries which refer to equations of the type

$$(f_{x_i})_{\overline{x_j}} = (f_{\overline{x_i}})_{x_j},$$

we speak of negative symmetries if conditions of the form

$$(f_{x_i})_{x_j} = (f_{\overline{x_i}})_{\overline{x_j}}$$

are satisfied.

9.3 Quantitative Statements

In this section the optimization quality of the presented ordering strategies is compared by means of some experimental results. For this purpose, we consider the circuits from the ISCAS '85 benchmark set. This set of 10 combinational circuits was published on the International Symposium on Circuits and Systems (ISCAS) in 1985 and was intended to compare experimental results in the field of testing and testability of circuits. Due to their wide distribution they have also been used in other areas. Here, we would like to point out that benchmark sets are necessary in many fields in order to be able to compare experimental results of different algorithms. However, one should always keep in mind that *no* set of benchmarks can represent the set of practically important function *completely*. Nevertheless, the ISCAS '85 set is suitable for our purposes to illustrate some typical effects.

The 10 circuits represent typical functionalities in the following fields of application:

- arithmetic logical units (ALUs),
- control units,
- selector units,
- priority decoding,
- error-correcting codes.

One of the 10 circuits is a 16-bit multiplier. Due to the exponential lower bound for multipliers proven in Section 8.2 this circuit cannot be represented by an OBDD with reasonable memory requirements. Therefore we will restrict the following discussion to the 9 remaining circuits.

The circuits are given in form of net lists of gates. In the subsequent comparison we would like to compare the fan-in heuristic, the window permutation algorithm, the sifting algorithm, and the symmetric sifting algorithm. For this reason, we consider the running times and the memory consumption in case of a symbolic simulation. The presented running times refer to a Sun

Name	In	Out	Gates	Successes	Space out	Time out	Size	Time
C432	36	7	160	10	0	0	490533	130.93
C499	41	32	202	10	0	0	840317	91.69
C880	60	26	383	10	0	0	776463	71.20
C1355	41	32	546	9	1	0	886374	198.09
C1908	33	25	880	10	0	0	68095	28.31
C2670	233	140	1193	0	9	1	–	–
C3540	50	22	1669	1	9	0	659838	258.66
C5315	178	32	2307	0	9	1	–	–
C7552	207	108	3512	0	9	1	–	–

Figure 9.15. Experimental results for random orders

Sparc 10 workstation. If the memory requirements exceed 100 MB or the symbolic simulation needs more than 20000 seconds, the process is interrupted (memory out or time out, respectively).

As starting point, we first consider the case of a random variable order. For each of the 9 investigated circuits, 10 random variable orders are determined. The table in Fig. 9.15 shows the running times of the symbolic simulation and the memory consumption of the OBDDs at the end of this process. The first four columns contain the names of the circuits, the number of input and output variables, and the number of gates of the circuits. Columns 5 to 7 tell us the number of successful computations, the number of memory overflows, and the number of time outs. In the last two columns the average number of nodes of the resulting OBDDs and the average running times are given. Here, the average values refer to the number of successful runs.

In particular, only for 4 of the 9 circuits did the symbolic simulation succeed for any of the 10 random orders. The resulting numbers of nodes lie in the range above half a million nodes with the exception of C1908.

The table in Fig. 9.16 shows the memory consumption of the resulting OBDDs when applying different algorithms for constructing good orders. Let us recall that different OBDD sizes for the same circuit *solely* go back to the fact that the corresponding algorithms lead to different orders and hence to different OBDD representations. The first column of the table contains the names of the circuit, the second and the third column the number of inputs and outputs. In the remaining columns we give the resulting OBDD sizes for the fan-in heuristic, the window permutation algorithm with window size 3 (window3), the window permutation algorithm with window size 4 (window4), the sifting algorithm, and the symmetric sifting algorithm. Figure 9.17 shows the required running times in the same way.

Of course, the exact numbers of the individual methods depend on implementation details, but the presented numbers reflect the trends quite well. Our experiments were conducted using the CUDD package, with the parameters set according to standard parameters. For example, the dynamic reordering

Name	In	Out	Fan-in	Window3	Window4	Sift	Symm.
C432	36	7	131178	1228	1226	1210	1210
C499	41	32	53866	39405	26541	26624	26624
C880	60	26	550302	14514	8388	10440	10440
C1355	41	32	53866	37470	26541	29562	29562
C1908	33	25	17758	21863	21587	6395	6395
C2670	233	140	Mem Out	Mem Out	Mem Out	4007	4007
C3540	50	22	Mem Out	471235	328242	23950	23950
C5315	178	32	Mem Out	Mem Out	Mem Out	1844	1844
C7552	207	108	Mem Out	Mem Out	Mem Out	8241	7895

Figure 9.16. OBDD sizes at the end of the symbolic simulation

algorithms are called whenever the size of an OBDD has doubled since the last reordering step.

First, by comparing the tables in Figs. 9.15 and 9.16 it can be seen that the presented optimization algorithms indeed substantially improve the situation for several instances. The resulting OBDDs are often much smaller. In general, the larger and the more complex the relevant functions are, the more extreme this effects becomes.

The two tables in Figs. 9.16 and 9.17 tell us that the dynamic reordering algorithms are clearly superior to the fan-in heuristic. The representations become substantially smaller, and there fewer are circuits whose OBDD computation fails because of a memory overflow. However, cases like the circuits C1355 or C1908 also point out that dynamic reordering algorithms need much computation time.

Furthermore, the tables tell us that the window permutation algorithm with window size 4 typically optimizes the representation better than the algorithm with window size 3. On the other hand, the time consumption is much larger in case of window size 4.

In comparison to the other methods, the sifting algorithm as well as the symmetric sifting variant are obviously superior. Only these methods succeed in transforming all 9 circuits into an OBDD representation. The resulting OBDD sizes are never more than slightly worse compared to the results of other strategies, and particularly for large circuits, the resulting OBDDs are much better. Although the effort required for a single application of sifting is generally bigger than that for a single application of the window permutation algorithm, the sifting-based symbolic simulation is much faster for the circuit C3540. This can be explained by the fact that during the whole time much smaller representations are involved.

Only in one instance of the mentioned examples did the symmetric sifting algorithm produce a better result than the sifting algorithm. This is because the rubber band effect from Section 9.2.5 occurs in none of the circuits in an extreme way. However, there are other real-world circuits where the influence

Name	In	Out	Fan-in	Window3	Window4	Sift	Symm.
C432	36	7	10	1	2	2	2
C499	41	32	5	14	30	77	74
C880	60	26	107	11	21	40	39
C1355	41	32	11	43	100	338	311
C1908	33	25	5	22	66	27	26
C2670	233	140	Mem Out	Mem Out	Mem Out	42	42
C3540	50	22	Mem Out	710	2408	175	175
C5315	178	32	Mem Out	Mem Out	Mem Out	17	17
C7552	207	108	Mem Out	Mem Out	Mem Out	105	106

Figure 9.17. Running times of the symbolic simulation

of this effect is quite strong. For this reason, the symmetric sifting algorithm typically works more stably than the sifting algorithm.

9.4 Outlook

Although the presented techniques work effectively in many applications, the design of good algorithms for optimizing the variable order remains an active research area. One of the still unsolved problems in the application of dynamic reordering originates from the fact that the presented methods, such as the sifting algorithm, require extremely much computation time if very large OBDDs with more than a million nodes are to be optimized. If the achieved size reduction is the deciding factor whether or not a computation succeeds, than the user is typically willing to accept this computation time. However, it is possible that at the end of a rather long sifting process we obtain the information: "Exactly 5 nodes have been gained."

One approach for solving this problem is to develop criteria which use structural information about the OBDDs in order to restrict the sifting process to suitable parts of the OBDD. Those parts in which no significant size reduction can be expected should not be considered. This strategy may completely eliminate the computation time of a searching process that is doomed to failure.

A first step in this direction was recently taken by Meinel and Slobodová. The method of **block-restricted sifting** picks up arguments from communication complexity that were used in the proofs of the lower bounds. These arguments allow one to compute promising parts within the OBDD. For each $1 \leq i \leq n$ the number of subfunctions $sf[i]$ is determined which result from fixing the first $i - 1$ variables in the order π to constants. The values $sf[i]$ can be interpreted as a function $sf : \{1, \ldots, n\} \to \mathbb{N}_0$ which is denoted as subfunction profile. A local minimum at a position i_0 of the subfunction profile indicates a weak information flow between the part of the OBDD above the i-th variable and the part below this variable. Hence, it does not seem to be reasonable to move variables from one of these parts into another one. The idea of block-restricted sifting is to extract sufficiently distinct local minima in the profile sf of subfunctions. Then the sifting algorithm is called for each sequence of variables which are located between two neighboring minima. During the sifting process no variable is moved across the border between different blocks, so much computation time can be gained, while the optimization quality of block-restricted sifting is only slightly worse than the quality of the original sifting algorithm.

9.5 References

The fan-in heuristic was proposed in the paper [MWBS88], the weight heuristic in [MIY90]. The first paper describing an efficient implementation of the variable swap is [FMK91]. The algorithm for exact minimization goes back to [FS90], its adaption to the framework of efficient variable swaps to [ISY91]. In the same paper [ISY91], the window permutation algorithm was presented.

In 1993, Rudell proposed the sifting algorithm [Rud93], while the symmetric variant is due to to Panda and Somenzi [PS95]. Finally, block-restricted sifting is described in the paper [MS97].

Part III

Applications and Extensions

10. Analysis of Sequential Systems

Cause célèbre [A sensational process].
François Gayot de Pitaval (1673–1743)

The design of increasingly complex electronic systems makes it more and more difficult to verify their correct behavior. At the same time it becomes more and more important that the systems work correctly, as nowadays human lives seriously depend on them, e.g., in traffic or in medicine.

The dramatic economic effects that errors in circuit design may cause is illustrated by the example of the Intel Pentium processor from the year 1994. In case of the Pentium implementation, a table of the SRT[1] divider circuit, which has been known and used for many years was set with incorrect entries. As a consequence, the processor – at least in some quite specific cases – did not compute the correct results. Although Intel argued for a long time that in practice this error would not have a serious influence on computations, the sense of uncertainty among affected PC users caused such a great public pressure that finally, a recall offer became unavoidable. The costs of this recall were estimated to 475 million US dollars. The lesson drawn from this debacle have had the effect that the field of *hardware verification* has become one of the essential steps within the design process.

Of particular importance is the verification of sequential systems, as any circuit (including combinational ones) can be modeled on the logic level by a finite state machine. The same holds true for the mentioned SRT divider circuit: it is nothing else but a finite state machine. In this chapter, we treat tools for the efficient analysis and verification of sequential systems. In particular, the paradigmatic application of the equivalence test presented in Section 5.3 turns out to be the core problem in quite different questions. Hence, we explain the individual techniques and their interplay in the context of this application.

---

[1] named after the initials of the three inventors

10.1 Formal Verification

We consider the following **general verification problem**:

Given: Two sequential systems: more precisely, two finite state machines M_1 and M_2 with the same number of input and output bits which are given by net lists of gates.

Question: Do M_1 and M_2 have the same input/output behavior, i.e., do M_1 and M_2 produce for each input sequence the same output sequence ?

The central idea of the OBDD-based solution approach is to reduce the verification of global properties to the verification of local properties which hold true for all states that can be reached from the initial state. For this reason, we first consider a seemingly simpler **restricted verification problem**:

Given: A finite state machine M, given by a net list of gates, with a single output $\lambda(x, e)$ over the output alphabet $\{0, 1\}$.

Question: Does M always produce the output value 1 for each possible input sequence ?

The restricted verification problem can be reduced to the verification of local properties by means of a reachability analysis. The term **reachability analysis** denotes the efficient computation and compact representation of all states which can be reached from the initial state. The characteristic function χ_R of this set $R \subset \mathbb{B}^n$ is a switching function and can therefore be represented by an OBDD (see Chapter 5).

If the characteristic function χ_R of the reachable states has been computed, then the restricted verification problem can be solved by checking the following simple Boolean equation:

$$(\chi_R(x_1, \ldots, x_n) \Rightarrow \lambda(x_1, \ldots, x_n, e_1, \ldots, e_p)) = 1 .$$

This equation can be rewritten as:

$$\overline{\chi_R(x_1, \ldots, x_n)} + \lambda(x_1, \ldots, x_n, e_1, \ldots, e_p) = 1 .$$

By constructing the so-called *product machine*, the task of solving the general verification problem, i.e., the equivalence test for two finite state machines, can be reduced to the restricted verification problem with respect to a single machine.

Definition 10.1. Let $M_1 = (Q_1, I, O, \delta_1, \lambda_1, q_1)$ and $M_2 = (Q_2, I, O, \delta_2, \lambda_2, q_2)$ be two finite state machines with p input bits and m output bits. The **product machine** $M = (Q, I, O, \delta, \lambda, q_0)$ of M_1 and M_2 is defined by

- $Q = Q_1 \times Q_2$,
- $\delta((x_1, x_2), e) = (\delta_1(x_1, e), \delta_2(x_2, e))$,
- $\lambda((x_1, x_2), e) = (\lambda_{1,1}(x_1, e) \equiv \lambda_{2,1}(x_2, e)) \cdot \ldots \cdot (\lambda_{1,m}(x_1, e) \equiv \lambda_{2,m}(x_2, e))$,
- $q_0 = (q_1, q_2)$.

The product machine in Fig. 10.1 simulates the behavior of M_1 and M_2, and produces the output 1 whenever the outputs of M_1 and M_2 coincide.

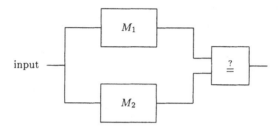

Figure 10.1. Schematic view of a product machine

Hence, reachability analysis plays a central role in formal verification of sequential systems. In the next section, we first treat several operators which turn out to be quite useful in this context. Then, in the subsequent sections, we show how these tools can be employed for efficiently realizing OBDD-based reachability analysis.

10.2 Basic Operators

10.2.1 Generalized Cofactors

The efficient computation of cofactors provides a key concept for performing many operations on OBDDs. Here, Shannon's expansion *with respect to the literals x_i and $\overline{x_i}$* is used to decompose a function f into $f = x_i f_{x_i} + \overline{x_i} f_{\overline{x_i}}$. The literals x_i and $\overline{x_i}$ are quite specific functions, and the question arises of how far cofactors can also be defined with respect to more general functions. As it will turn out, the resulting concept is a central ingredient in the field of image computation.

Definition 10.2. *A set of switching functions $\{b_1, \ldots, b_r\}$, $b_i \in \mathbb{B}_n$, is called* **orthonormal**, *if*

- $\sum_{i=1}^{r} b_i = 1$, *and*
- $b_i \cdot b_j = 0$ *for $1 \leq i, j \leq r$ and $i \neq j$.*

Example 10.3. (1) For each variable x_i the set $\{x_i, \overline{x_i}\}$ is an orthonormal set.

(2) For each function $f \in \mathbb{B}_n$ the set $\{f, \overline{f}\}$ is orthonormal.   ◇

For each orthonormal set $\{b_1, \ldots, b_r\}$ a function f can be written in the form

$$f(x_1, \ldots, x_n) = \sum_{i=1}^{r} f_i(x_1, \ldots, x_n) \cdot b_i(x_1, \ldots, x_n).$$

Here, the coefficients f_i depend on f and on $\{b_1, \ldots, b_r\}$. They are characterized by the following result.

Theorem 10.4. *Let $\{b_1, \ldots, b_r\}$ be an orthonormal set, and let $f \in \mathbb{B}_n$. The decomposition*

$$f = \sum_{i=1}^{r} f_i b_i$$

holds true if and only if all the functions f_i satisfy

$$f \cdot b_i = f_i \cdot b_i \quad \text{for all } 1 \le i \le r.$$

Proof. First let $f \cdot b_i = f_i \cdot b_i$ for all $1 \le i \le r$. Then

$$\sum_{i=1}^{r} f_i \cdot b_i = \sum_{i=1}^{r} f \cdot b_i = f \cdot \sum_{i=1}^{r} b_i = f.$$

If, conversely, $f = \sum_{j=1}^{r} f_j b_j$, then

$$f \cdot b_i = (\sum_{j=1}^{r} f_j \cdot b_j) \cdot b_i = f_i \cdot b_i$$

due to the second basic property of orthonormal functions.   □

In general, within the decomposition of Theorem 10.4 the coefficients f_i are not uniquely determined. Indeed, for an element b_i of the orthonormal set each coefficient f_i with $f \cdot b_i \le f_i \le f + \overline{b_i}$ satisfies the condition $f \cdot b_i = f_i \cdot b_i$. This follows from

$$f_i = f \cdot b_i \implies f_i \cdot b_i = f \cdot b_i \cdot b_i = f \cdot b_i,$$
$$f_i = f + \overline{b_i} \implies f_i \cdot b_i = (f + \overline{b_i}) \cdot b_i = f \cdot b_i + \overline{b_i} \cdot b_i = f \cdot b_i.$$

This observation can be interpreted in the way that f_i has to coincide with f at least at those positions where $b_i = 1$. At the positions with $b_i = 0$ the function f_i can be chosen arbitrarily. In the later discussion of image computation this freedom will be used to obtain more compact representations of the relevant functions. The next theorem implies that many properties of usual cofactors are also valid in the more general context.

Theorem 10.5. *For an orthonormal set $\{b_1, \ldots, b_r\}$ and two switching functions $f = \sum_{i=1}^{r} f_i b_i$, $g = \sum_{i=1}^{r} g_i b_i$ we have*

$$f + g = \sum_{i=1}^{r} (f_i + g_i) b_i,$$

$$f \cdot g = \sum_{i=1}^{r} (f_i \cdot g_i) b_i,$$

$$\overline{f} = \sum_{i=1}^{r} \overline{f_i} b_i.$$

Proof. The proof of the first statement follows immediately from the distributive laws. The second statement follows from $b_i \cdot b_j = 0$ for $i \neq j$. The proof of the third property is based on the complete characterization of the complement via

$$f \cdot \overline{f} = 0 \quad \text{and} \quad f + \overline{f} = 1,$$

a fact that has already been stated in the proof of Theorem 3.5.

By substituting the expressions of f and $\overline{f}$, we obtain

$$f \cdot \overline{f} = \left(\sum_{i=1}^{r} f_i b_i\right) \cdot \left(\sum_{i=1}^{r} \overline{f_i} b_i\right) = \sum_{i=1}^{r} (f_i \cdot \overline{f_i}) b_i = 0$$

and

$$f + \overline{f} = \left(\sum_{i=1}^{r} f_i b_i\right) + \left(\sum_{i=1}^{r} \overline{f_i} b_i\right) = \sum_{i=1}^{r} (f_i + \overline{f_i}) b_i = \sum_{i=1}^{r} b_i = 1.$$

$\square$

Concerning the decomposition of a function with respect to the orthonormal set $\{g, \overline{g}\}$ for a $g \in \mathbb{B}_n$, the analogy to Shannon's decomposition of this function suggests the notion of a *generalized cofactor*.

Definition 10.6. *Let $f, g \in \mathbb{B}_n$, and let*

$$f = g \cdot f_g + \overline{g} \cdot f_{\overline{g}}$$

*be a decomposition of f with respect to the orthonormal set $\{g, \overline{g}\}$. Then the coefficient f_g is called **positive generalized cofactor** of f with respect to g, and the coefficient $f_{\overline{g}}$ is called **negative generalized cofactor** of f with respect to g.*

Lemma 10.7. *Let $f, g \in \mathbb{B}_n$ with $f \cdot g = 0$, and let $f_{\overline{g}}$ be a negative generalized cofactor with respect to g. Then*

$$f \leq f_{\overline{g}} \leq f + g.$$

Proof. The equation $f \cdot g = 0$ implies $f = \bar{g} \cdot f_{\bar{g}}$. Then the claim follows from

$$f = \bar{g} f_{\bar{g}} \leq f_{\bar{g}},$$
$$f + g = \bar{g} f_{\bar{g}} + g \cdot (f_{\bar{g}} + \overline{f_{\bar{g}}}) = f_{\bar{g}} + g \cdot \overline{f_{\bar{g}}} \geq f_{\bar{g}}.$$

$\square$

10.2.2 The Constrain Operator

As already mentioned, usually generalized cofactors of a function f with respect to a function g are not uniquely determined. For this reason, it is possible to construct well-suited cofactors with respect to a given optimization criterion. The typical criterion in our context is the OBDD size of the cofactor.

In the previous section, we have noticed that the values of the coefficients f_g and $f_{\bar{g}}$ in the generalized cofactor decomposition

$$f = g \, f_g + \bar{g} \, f_{\bar{g}}$$

are uniquely determined for the positions $(x_1, \dots, x_n)$ with $g(x_1, \dots, x_n) = 1$ and $\bar{g}(x_1, \dots, x_n) = 1$, respectively. However, for the positions $(x_1, \dots, x_n)$ with $g(x_1, \dots, x_n) = 0$, respectively $\bar{g}(x_1, \dots, x_n) = 0$, there is some freedom of choice.

The *constrain operator* defined below serves for the computation of compact cofactors. We restrict to the consideration of positive cofactors, the results can also be immediately transferred to negative cofactors. The idea of the constrain operator is to map each minterm in the off-set of g to a minterm in the on-set of g. This mapping is used in order to choose the values of f_g for the positions $(x_1, \dots, x_n)$ with $g(x_1, \dots, x_n) = 0$. The mapping is based on the following definition of the *distance* of two input vectors.

Definition 10.8. *Let the variables $x_1, \dots, x_n$ be ordered in the order π according to $x_{j_1} < x_{j_2} < \dots < x_{j_n}$. Let $r = (r_1, \dots, r_n)$, $s = (s_1, \dots, s_n) \in \mathbb{B}^n$. The **distance** $\|r - s\|$ of r and s **with respect to the order π** is defined by*

$$\|r - s\| = \sum_{i=1}^{n} |r_{j_i} - s_{j_i}| \, 2^{n-i}.$$

Definition 10.9. *For $f, g \in \mathbb{B}_n$ the **constrain operator** $f \downarrow g$ is defined by*

$$(f \downarrow g)(r) = \begin{cases} f(r) & \text{if } g(r) = 1, \\ f(s) & \text{if } g(r) = 0, \ g(s) = 1 \text{ and } \|r - s\| \text{ minimal}, \\ 0 & \text{if } g = 0. \end{cases}$$

```
constrain(f, g) {
/* Input: OBDDs of f, g ∈ 𝔹ₙ */
/* Output: An OBDD of f ↓ g */
    Let xᵢ be the top variable in {f, g};
    If (g = 1 or f = 0 or f = 1) Return f;
    Else If (f = g) Return 1;
    Else If (f = ḡ) Return 0;
    Else If (g = 0) Return 0;
    Else If (g_{xᵢ} = 0) Return constrain(f_{x̄ᵢ}, g_{x̄ᵢ});
    Else If (g_{x̄ᵢ} = 0) Return constrain(f_{xᵢ}, g_{xᵢ});
    Else Return ITE(xᵢ, constrain(f_{xᵢ}, g_{xᵢ}), constrain(f_{x̄ᵢ}, g_{x̄ᵢ}));
}
```

Figure 10.2. Constrain algorithm

As at all positions $(x_1, \ldots, x_n)$ with $g(x_1, \ldots, x_n) = 1$ the function $f \downarrow g$ coincides with f, we observe that $f \downarrow g$ is a positive generalized cofactor of f with respect to g.

Figure 10.2 contains pseudo code for computing the constrain operator. The algorithm can be implemented efficiently by using the caching techniques from Chapter 7. The proof that the presented algorithm computes the constrain operator correctly is based on the following theorem.

Theorem 10.10. *Let* $f, g, h \in \mathbb{B}_n$, *and let*

$$\tilde{f}_g = h \cdot (f_h)_{g_h} + \overline{h} \cdot (f_{\overline{h}})_{g_{\overline{h}}} \quad and \quad \tilde{f}_{\overline{g}} = h \cdot (f_h)_{\overline{g}_h} + \overline{h} \cdot (f_{\overline{h}})_{\overline{g}_{\overline{h}}}.$$

Then

$$f = g \cdot \tilde{f}_g + \overline{g} \tilde{f}_{\overline{g}}.$$

Proof. We show that the condition in Theorem 10.4 is satisfied for the two functions $\tilde{f}_g$ and $\tilde{f}_{\overline{g}}$. For $\tilde{f}_g$ we have

$$\begin{aligned}
\tilde{f}_g \cdot g &= g \cdot h \cdot (f_h)_{g_h} + g \cdot \overline{h} \cdot (f_{\overline{h}})_{g_{\overline{h}}} \\
&= h \cdot g_h \cdot (f_h)_{g_h} + \overline{h} \cdot g_{\overline{h}} \cdot (f_{\overline{h}})_{g_{\overline{h}}} \\
&= h \cdot g_h \cdot f_h + \overline{h} \cdot g_{\overline{h}} \cdot f_{\overline{h}} \\
&= h \cdot g_h \cdot f + \overline{h} \cdot g_{\overline{h}} \cdot f \\
&= f \cdot (h \cdot g + \overline{h} \cdot g) = f \cdot g.
\end{aligned}$$

Analogously, $\tilde{f}_{\overline{g}} \cdot \overline{g} = f \cdot \overline{g}$ can be proven, and this concludes the proof.   □

Theorem 10.11. *The algorithm in Fig. 10.2 computes the constrain operator* $f \downarrow g$.

Proof. The proof of the statements is carried out by means of a case distinction. To simplify the notation we assume that π is the natural variable order $x_1 < x_2 < \ldots < x_n$. The terminal cases are

$g = 1$: In this case we have $f \downarrow g = f$ in agreement with the algorithm.

f constant: We have $f \downarrow g = f$ in agreement with the algorithm.

$f = g$: For all s with $g(s) = 1$ we have $f(s) = 1$. This implies $f \downarrow g = 1$.

$f = \overline{g}$: For all s with $g(s) = 1$ we have $f(s) = 0$. This implies $f \downarrow g = 0$.

$g = 0$: The definition of the constrain operator implies $f \downarrow g = 0$.

The recursive cases are

$g_{x_i} = 0$: Let $c \in \mathbb{B}^p$ be the part of the vector which contains the already fixed variables $x_1, \ldots, x_p$. The task is to find the vector s with $g(s) = 1$ which has minimal distance to $(c, 1, \sqcap\_\sqcap) = (c_1, \ldots, c_p, 1, \sqcap\_\sqcap)$, where $\sqcap\_\sqcap$ symbolizes the still unspecified variables.

First we remark that this vector has to be contained in the set which is defined by $(c, 0, \sqcap\_\sqcap)$. This holds true, as the weights in the definition of the distance decrease exponentially from the root to the sinks. Hence, each vector in $(c, 0, \sqcap\_\sqcap)$ is nearer to the vectors in $(c, 1, \sqcap\_\sqcap)$ than any other relevant vector.

Now we consider a vector $r = (c, 1, t)$ in the set $(c, 1, \sqcap\_\sqcap)$. If the vector $\tilde{r} = (c, 0, t)$ implies $g(\tilde{r}) = 1$ then $s = \tilde{r}$. In the other case, s is the vector in $\mathbb{B}^n$ with $g(s) = 1$ which has the smallest distance from $\tilde{r}$. This is exactly what the algorithm computes.

$g_{\overline{x_i}} = 0$: Analogous to the case $g_{x_i} = 0$.

Otherwise: In this case a recursive decomposition according to Theorem 10.10 is performed. $\qquad\square$

Example 10.12. An example for the application of the constrain operator is given in Fig. 10.3. The functions f and g are $f(x) = x_1 x_3 + \overline{x_1}(x_2 \oplus x_3)$ and $g(x) = x_1 x_2 + \overline{x_2}\ \overline{x_3}$. As the algorithm exploits the property that complemented edges allow to check the condition $f = \overline{g}$ in constant time, the OBDDs shown include complemented edges. For $h = f \downarrow g$ we have

$$
\begin{aligned}
h = f \downarrow g &= x_1 \cdot (f_{x_1} \downarrow g_{x_1}) + \overline{x_1} \cdot (f_{\overline{x_1}} \downarrow g_{\overline{x_1}}) \\
&= x_1 \cdot (x_3 \downarrow q) + \overline{x_1} \cdot (\overline{p} \downarrow \overline{r}) \\
&= x_1 \cdot (x_2 \cdot (x_3 \downarrow 1) + \overline{x_2} \cdot (x_3 \downarrow \overline{x_3})) + \overline{x_1} \cdot (x_3 \downarrow \overline{x_3}) \\
&= x_1 \cdot (x_2 \cdot x_3 + \overline{x_2} \cdot 0) + \overline{x_1} \cdot 0 \\
&= x_1 x_2 x_3.
\end{aligned}
$$

The truth table of f, g, and $f \downarrow g$ in Fig. 10.3 illustrates the minimization idea of the constrain operator. For input vectors $r \in \mathbb{B}^n$ with $g(r) = 0$ the idea is to minimize the OBDD by taking over the values $f(s)$ from close vectors s. $\qquad\diamond$

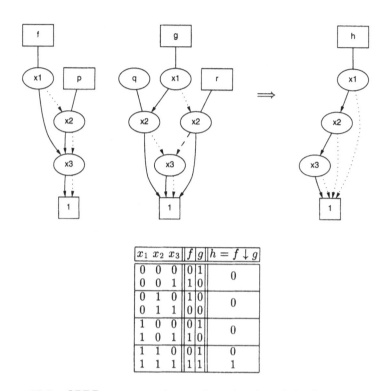

x_1 x_2 x_3	f	g	$h = f \downarrow g$
0 0 0	0	1	0
0 0 1	1	0	
0 1 0	1	0	0
0 1 1	0	0	
1 0 0	0	1	0
1 0 1	1	0	
1 1 0	0	1	0
1 1 1	1	1	1

Figure 10.3. OBDD representations and truth table of the functions f, g and $h = f \downarrow g$

We would like to conclude the discussion of the constrain operator by pointing out a difficulty. Although the OBDD of $f \downarrow g$ usually contains fewer nodes than the OBDD of f, there are cases where the opposite can be observed. In particular, this effect may occur if the OBDD of g is large and depends essentially on many variables which are not essential for f. These variables can occur in $f \downarrow g$ and hence cause an undesirable growth of the graphs.

In the context of cofactor computation this problem can be solved by quantification and application of the restrict operator, two concepts which will be treated in the next two sections.

10.2.3 Quantification

The *Boolean existential quantifier* and *Boolean universal quantifier* defined below are operators which turn out to be quite essential in the context of sequential analysis.

Definition 10.13. *For $f \in \mathbb{B}_n$ the **existential quantification** with respect to the variable x_i is defined by*

$$\exists_{x_i} f = f_{x_i} + f_{\overline{x_i}}.$$

*The **universal quantification** with respect to x_i is defined by*

$$\forall_{x_i} f = f_{x_i} \cdot f_{\overline{x_i}}.$$

Both $\exists_{x_i} f$ and $\forall_{x_i} f$ denote switching functions which no longer depend on the variable x_i. Of course, the notion of quantification originates from the equivalences

$$(\exists_{x_i} f)(x_1, \ldots, x_{i-1}, x_{i+1}, \ldots, x_n) = 1 \iff \exists_{x_i} (f(x_1, \ldots, x_n) = 1),$$
$$(\forall_{x_i} f)(x_1, \ldots, x_{i-1}, x_{i+1}, \ldots, x_n) = 1 \iff \forall_{x_i} (f(x_1, \ldots, x_n) = 1),$$

where $\forall$ and $\exists$ on the right side of both equivalences denote the usual quantifiers from predicate calculus.

Example 10.14. Let $f(x_1, x_2, x_3) = \overline{x_1}\,\overline{x_2}x_3 + x_1\overline{x_3} + x_1x_2$. Then the two cofactors with respect to x_3 are

$$f_{x_3} = \overline{x_1}\,\overline{x_2} + x_1x_2 \quad \text{and} \quad f_{\overline{x_3}} = x_1 + x_1x_2 = x_1,$$

and hence,

$$\exists_{x_3} f = x_1 + \overline{x_2} \quad \text{and} \quad \forall_{x_3} f = x_1x_2.$$

$\diamond$

If one considers $\exists_{x_i} f$ and $\forall_{x_i} f$ still as functions in $\mathbb{B}_n$, then it can be stated: $\exists_{x_i}$ is the smallest function whose on-set is contained in the on-set of f and which is independent of x_i. $\forall_{x_i}$ is the largest function whose on-set contains the on-set of f and which is independent of x_i.

Lemma 10.15. *For $f, g \in \mathbb{B}_n$ the following properties of quantification functions can be stated:*

Monotony:

$$f \leq g \implies \exists_{x_i} f \leq \exists_{x_i} g \quad \text{and} \quad \forall_{x_i} f \leq \forall_{x_i} g.$$

Commutativity:

$$\exists_{x_i} \exists_{x_j} f = \exists_{x_j} \exists_{x_i} f \quad \text{and} \quad \forall_{x_i} \forall_{x_j} f = \forall_{x_j} \forall_{x_i} f.$$

Distributivity:

$$\exists_{x_i} (f + g) = \exists_{x_i} f + \exists_{x_i} g \quad \text{and} \quad \forall_{x_i} (f \cdot g) = \forall_{x_i} f \cdot \forall_{x_i} g$$

Distributivity inequations:

$$\exists_{x_i}(f \cdot g) \le \exists_{x_i} f \cdot \exists_{x_i} g \quad and \quad \forall_{x_i}(f + g) \ge \forall_{x_i} f + \forall_{x_i} g$$

Complementation:

$$\exists_{x_i}(\overline{f}) = \overline{(\exists_{x_i} f)} \quad and \quad \forall_{x_i}(\overline{f}) = \overline{(\forall_{x_i} f)}$$

$\square$

The proof can easily be carried out by individually checking the equations. Due to commutativity, we can abbreviate $\exists_{x_1} \exists_{x_2} \ldots \exists_{x_n} f$ by $\exists_{x_1,x_2,\ldots,x_n}$ or for short by $\exists_x$.

10.2.4 The Restrict Operator

Lemma 10.16. *Let f, g, and h be switching functions with $h \ge g$. Then*

$$g \cdot f_g = g \cdot f_h.$$

In other words: a positive generalized cofactor with respect to h is also a positive generalized cofactor with respect to g.

Proof. A positive generalized cofactor of f with respect to g has to coincide with f at all positions x with $g(x) = 1$. This coincidence also applies to f_h, as $g(x) = 1$ implies $h(x) = 1$. $\square$

In particular, the lemma can be applied to the constrain operator from Section 10.2.2 and to $h = \exists_{x_i} g$. In this way, the variables of g which do not occur in f and which may cause an undesirable growth in size during the computation of the constrain operator can be eliminated by existential quantification before the computation of $f \downarrow g$. If h denotes the result of this quantification, then $f \downarrow h$ is also a cofactor of f with respect to g.

As a consequence, quantification can be integrated into the computation of the constrain operator. The operator computed by this modified algorithm is called the **restrict operator** and is denoted by $f \Downarrow g$. Whenever the top variable x_i in g has a smaller index than the top variable in f, then the function returns the result

$$f \Downarrow (\exists_{x_i} g).$$

The algorithm for computing the restrict operator is shown in Fig. 10.4. It originates from the constrain algorithm in Fig. 10.2 by the incorporation of the existential quantification.

In particular in the cases of cofactor computations where the OBDD is large and depends on many variables, the restrict operator is superior to the constrain operator: typically, the generated OBDDs are significantly more compact.

```
restrict(f, g) {
/* Input: OBDDs of f, g ∈ Bₙ */
/* Output: An OBDD of f ⇓ g */
    Let xᵢ be the top variable {f, g};
    If (g = 1 or f = 0 or f = 1) Return f;
    Else If (f = g) Return 1;
    Else If (f = g̅) Return 0;
    Else If (g = 0) Return 0;
    Else If (gₓᵢ = 0) Return restrict(fₓ̄ᵢ, gₓ̄ᵢ);
    Else If (gₓ̄ᵢ = 0) Return restrict(fₓᵢ, gₓᵢ);
    Else If (xᵢ is not the top variable in f) {
        Return restrict(f, ITE(gₓᵢ, 1, gₓ̄ᵢ));
    }
    Else Return ITE(xᵢ, restrict(fₓᵢ, gₓᵢ), restrict(fₓ̄ᵢ, gₓ̄ᵢ));
}
```

Figure 10.4. Restrict algorithm

10.3 Reachability Analysis

Definition 10.17. *Let $M = (Q, I, O, \delta, \lambda, q_0)$ be a finite state machine. A state $s \in \mathbb{B}^n$ is said to be* **reachable in exactly k steps from the state** r *if there is an input sequence $e_0, \ldots, e_{k-1}$ and a state sequence $s_0, \ldots, s_k$ such that $s_0 = r$, $s_k = s$ and*

$$\delta(s_i, e_i) = s_{i+1}, \qquad 0 \leq i < k.$$

s is said to be **reachable** *from the state r if there is an integer $k \geq 0$ such that s is reachable from r in exactly k steps.*

The term *reachability analysis* of a finite state machine M denotes the computation and efficient representation of all states which are reachable from the initial state. In our context, the relevant state sets are represented in terms of OBDDs of their characteristic function.

For a finite state machine M with p input bits, n state bits and next-state function $\delta : \mathbb{B}^{n+p} \to \mathbb{B}^n$, let $\chi_j(x_1, \ldots, x_n) : \mathbb{B}^n \to \mathbb{B}$ denote the characteristic function of all states which are reachable in at most j steps.

Definition 10.18. *Let $f : \mathbb{B}^n \to \mathbb{B}^m$. The* **image** *$Im(f)$ of the function f is defined by*

$$Im(f) = \{v \in \mathbb{B}^m : \text{ there exists some } x \in \mathbb{B}^n \text{ with } f(x) = v\}.$$

For a subset C of $\mathbb{B}^n$ the **image of f with respect to C** *is defined by*

$$Im(f, C) = \{v \in \mathbb{B}^m : \text{ there exists some } x \in C \text{ with } f(x) = v\}.$$

```
traverse(δ, q₀) {
/* Input: Next-state function δ, initial set S₀ */
/* Output: Set of reachable states */
    Reached = From = S₀;
    Do {
        To = Im(δ, From);
        New = To \ Reached;
        From = New;
        Reached = Reached ∪ New;
    } While (New ≠ ∅);
    Return Reached;
}
```

Figure 10.5. Basic algorithm for reachability analysis based on breadth-first traversal

The basic structure of the algorithm for computing the set *Reached* of reachable states in M is shown in Fig. 10.5. The algorithm is based on breadth-first-traversal. First, the set *To* of all successor states of the initial set S_0 is computed. In order to keep all the OBDDs occurring during the image operation small, the subset *New* $\subset$ *To* is computed which filters all states that are already contained in the starting set. The set *From*, which is the starting point of the next image computation, is set to *New*, and the set *Reached* of all previously computed states is updated. Then the next iteration begins.

The iteration can be stopped if during one step no new states are added. The set union and the construction of the set difference are performed by means of the corresponding Boolean operations on the OBDDs of the characteristic functions. The number of totally performed iterations corresponds to the **sequential depth** of the finite state machine. This term denotes the smallest integer $k \geq 0$ such that each state is reachable from the initial state in at most k steps.

The computation of the image operator is a rather costly operation, we will explain algorithms for performing this operation in the next section. Independently of the choice of a particular algorithm, it is reasonable and necessary during the computation of $Im(\delta, From)$ to keep the OBDD of the input set *From* as simple as possible. For this reason, we have demanded *From* = *New* during the traversal, i.e., exactly the newly computed states constitute the starting point of the new computation.

Obviously, instead of *From* = *New*, any other set *From* with the property

$$New \subset From \subset Reached \cup New$$

can also be used. According to Lemma 10.7, each generalized cofactor $(New)_{\overline{Reached}}$ satisfies this property:

$$New \subset (New)\overline{_{Reached}} \subset Reached \cup New.$$

Hence, it is wiser not to set *From = New*, but to use the restrict operator from Section 10.2.4 to compute a suitable *From*. Namely, this operator computes a generalized cofactor whose OBDD size is rather small. For this reason, we choose as starting point for the new computation:

$$From = New \Downarrow \overline{Reached}.$$

Here, we remark again that in an explicit representation of the states each state has to be touched explicitly. By contrast, in an implicit representation by means of the characteristic function, the image can be computed by performing a single operator, the image operator. Hence, one also speaks of **symbolic breadth-first traversal**. Here, the difficulty in handling large numbers of states is again shifted to the size of the employed OBDD representation.

However, the image operator constitutes a far more complex operation than all the operations introduced earlier. Hence, in the next section, we shall be intensively concerned with good algorithms for computing this important operator. But here we would already like to mention that the algorithms for dynamic reordering presented in Chapter 9 are of central importance in sequential analysis. Although heuristic methods are able to construct good orders for the first steps of an iteration process within reachability analysis, the dynamic behavior of the iteration process cannot be captured. Instead, dynamic reordering algorithms allow the order to be adapted to the varying state sets continuously. In the presence of dynamic reordering, a large fraction of the running time is put into the construction of good orders. However, this time is invested well, as this may be the factor deciding whether or not a computation succeeds.

10.4 Efficient Image Computation

The central operation for reachability analysis is the **image computation** with respect to a subset C. In general, computing the image operator is substantially more complex than, say, computing binary operations. If during the performance of an image computation the variable order is kept constant, then, of course, the OBDD size of the image is independent of the algorithm used for this computation. However, for all known algorithms of image computation, the effect illustrated in Fig. 10.6 may occur. Although the initial set C and the image $Im(f, C)$ are of acceptable size, the sizes of the intermediate results during the computation explode. Hence, good algorithms for image computation aim primarily at keeping possible intermediate results small.

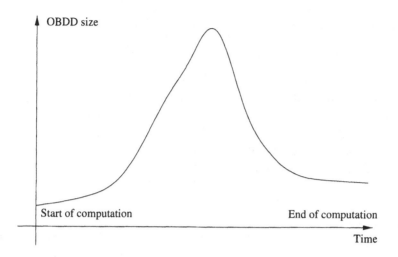

Figure 10.6. Explosion of memory consumption during image computation

In the following, all sets will be represented in terms of their characteristic function. Then, in case of *restricted image computation* the problem to be solved can be stated in the following way:

Given: Two OBDDs of the switching function $f(x_1, \ldots, x_n)$ and the characteristic function χ_C of a subset $C \subset \{0,1\}^n$.

Wanted: The OBDD of the characteristic function of $\mathrm{Im}(f, C)$.

Typically, we express the characteristic function of $\mathrm{Im}(f, C)$ in the variables $y_1, \ldots, y_m$.

10.4.1 Input Splitting

Let $f : \mathbb{B}^n \to \mathbb{B}^m$. One of the possible methods for computing the unrestricted image $\mathrm{Im}(f)$ is called **input splitting**. It is based on the observation that the image computation can be decomposed with respect to to the input variables,

$$\mathrm{Im}(f) = \mathrm{Im}(f)_{x_i} + \mathrm{Im}(f)_{\overline{x_i}}.$$

We assume that the characteristic function of the image set is expressed in the variables $y_1, \ldots, y_m$. A function $f = (f_1, \ldots, f_m)$ which is constant in all components, i.e., $f_i \in \{0, 1\}$, constitutes a terminal case:

$$\mathrm{Im}(f) = y_1^{f_1} \cdot \ldots \cdot y_m^{f_m}.$$

The efficiency of this approach can be improved by exploiting heuristics and sharper stopping criteria.

Decomposition in the direction of disjunct carrier. Let $f = (f_1, f_2) : \mathbb{B}^n \to \mathbb{B}^2$, where f_1 and f_2 depend on disjunct variable sets. Then

$$\text{Im}(f) = \text{Im}(f_1) \times \text{Im}(f_2).$$

Of course, this statement can be generalized to functions $f = (f_1, \ldots, f_m)$ which can be divided into blocks with disjoint variable sets. The choice of decomposition variables in the recursive partition is performed with the aim of obtaining blocks with disjoint carrier sets as soon as possible.

Identical and complementary components. Let $f = (f_1, f_2) : \mathbb{B}^n \to \mathbb{B}^2$ be a non-constant function with $f_1 = f_2$ or $f_1 = \overline{f_2}$. Then we have

$$\text{Im}(f) = y_1 \equiv y_2 \quad \text{or} \quad \text{Im}(f) = y_1 \oplus y_2.$$

This statement can also be generalized to functions $f = (f_1, \ldots, f_m)$. A criterion for the choice of decomposition variables is to aim at reducing the problem to a problem with identical or complementary components as soon as possible. Namely, the latter ones can be deleted but one, and it remains to solve a simplified problem. Later on, the deleted components are added again by using terms of the form $y_i \equiv y_j$ or $y_i \oplus y_j$.

Identical subproblems. As in the case of computing binary operations, a computed table with the results of already solved subproblems is constructed.

Example 10.19. Let $f = (f_1, f_2, f_3)$ be defined by

$$f_1 = x_1(x_2 + x_3),$$
$$f_2 = x_2(x_1 + x_3),$$
$$f_3 = x_3(x_1 + x_2).$$

The positive cofactors $(f_i)_{x_1}$ are

$$(f_1)_{x_1} = x_2 + x_3, \qquad (f_2)_{x_1} = x_2, \qquad (f_3)_{x_1} = x_3.$$

These functions do not form terminal cases yet. Another decomposition with respect to x_2 yields

$$(f_1)_{x_1 x_2} = 1, \qquad (f_2)_{x_1 x_2} = 1, \qquad (f_3)_{x_1 x_2} = x_3.$$

This implies $\text{Im}(f)_{x_1 x_2} = y_1 y_2$. The negative cofactors with respect to x_2 are

$$(f_1)_{x_1 \overline{x_2}} = x_3, \qquad (f_2)_{x_1 \overline{x_2}} = 0, \qquad (f_3)_{x_1 \overline{x_2}} = x_3.$$

This implies $\text{Im}(f)_{x_1 \overline{x_2}} = (y_1 y_3 + \overline{y_1}\,\overline{y_3})\overline{y_2}$, and hence

$$\text{Im}(f)_{x_1} = y_1 y_2 + y_1 \overline{y_2} y_3 + \overline{y_1}\,\overline{y_2}\,\overline{y_3}$$
$$= y_1 y_2 + y_1 y_3 + \overline{y_1}\,\overline{y_2}\,\overline{y_3}.$$

Within the computation of $\mathrm{Im}(f)_{\overline{x_1}}$ we obtain

$$(f_1)_{\overline{x_1}} = 0, \qquad (f_2)_{\overline{x_1}} = (f_3)_{\overline{x_1}} = x_2 x_3.$$

Eliminating the identical subproblem immediately results in

$$\mathrm{Im}(f)_{\overline{x_1}} = \overline{y_1}(y_2 y_3 + \overline{y_2}\ \overline{y_3}).$$

Finally, the disjunction of the two partial results yields

$$\begin{aligned} \mathrm{Im}(f) &= \mathrm{Im}(f)_{x_1} + \mathrm{Im}(f)_{\overline{x_1}} \\ &= y_1 y_2 + y_1 y_3 + y_2 y_3 + \overline{y_1}\ \overline{y_2}\ \overline{y_3}. \end{aligned}$$

$\diamond$

For the computation of a restricted image $\mathrm{Im}(f,C)$ with respect to a subset $C \subset \mathbb{B}^n$ the constrain operator can be employed once more. By using this operator, the function f can be modified to a function f' such that the computation of $\mathrm{Im}(f,C)$ is transformed to an unrestricted image computation $\mathrm{Im}(f')$.

Lemma 10.20. *Let $f : \mathbb{B}^n \to \mathbb{B}^m$, and let $\chi_C \in \mathbb{B}_n$ be the characteristic function of a set $C \subset \mathbb{B}^n$. Then we have*

$$\mathrm{Im}(f,C) = \mathrm{Im}(f \downarrow \chi_C) = \mathrm{Im}(f_1 \downarrow \chi_C, \dots, f_n \downarrow \chi_C).$$

Proof. As for $C = \emptyset$ the claim is obviously satisfied, we can assume that $C \neq \emptyset$. Each $x \in C$ satisfies $(f \downarrow \chi_C)(x) = f(x)$. This implies $\mathrm{Im}(f,C) \subset \mathrm{Im}(f \downarrow \chi_C)$. On the other hand, for each $x \notin C$ there exists an $s \in C$ with

$$\begin{aligned} (f \downarrow \chi_C)(x) &= (f_1 \downarrow \chi_C(x), \dots, f_m \downarrow \chi_C(x)) & (10.1) \\ &= (f_1 \downarrow \chi_C(s), \dots, f_m \downarrow \chi_C(s)) & (10.2) \\ &= (f \downarrow \chi_C)(s) \\ &= f(s). \end{aligned}$$

Hence, $\mathrm{Im}(f \downarrow \chi_C) \subset \mathrm{Im}(f,C)$. Note that for functions f with several outputs, the step from (10.1) to (10.2) only holds because the coordinate shift of the constrain operator $f \downarrow g$ merely depends on g. $\square$

Each operator $\circ$ which satisfies the property $\mathrm{Im}(f,C) = \mathrm{Im}(f \circ \chi_C)$ is called an **image restrictor**.

The effect of an image restrictor is illustrated in Fig. 10.7. Due to the modification of the function, x is not mapped to $f(x)$, but to $f'(x) \in \mathrm{Im}(f,C)$.

In contrast to the constrain operator $\downarrow$, the restrict operator $f \Downarrow g$ is not an image restrictor, because the coordinate shift of the restrict operator also depends on the function f itself.

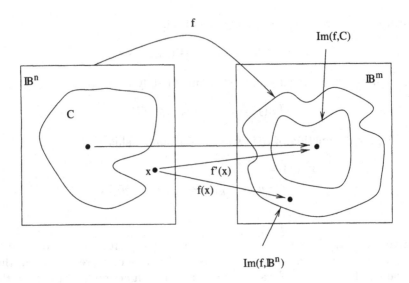

Figure 10.7. Effect of an image restrictor

10.4.2 Output Splitting

During image computation a function can be decomposed not only with respect to the input variables $x_1, \ldots, x_n$, but also with respect to the output variables $y_1, \ldots, y_m$:

$$\mathrm{Im}(f) = y_i \cdot \mathrm{Im}((f_1, \ldots, f_{i-1}, \cdot, f_{i+1}, \ldots, f_m), \mathrm{on}(f_i)) \qquad (10.3)$$
$$+ \, \overline{y_i} \cdot \mathrm{Im}((f_1, \ldots, f_{i-1}, \cdot, f_{i+1}, \ldots, f_m), \mathrm{on}(\overline{f_i})),$$

a method which is denoted as **output splitting**. Here, the dot serves to symbolize the currently missing i-th component. By decomposing according to (10.3), the image of f is divided into the elements with $f_i = 1$ and into the elements with $f_i = 0$. In this way, an unrestricted image computation is converted into the computation of two restricted images $\mathrm{Im}(f, C)$ of smaller dimension. By using the constrain operator and the technique presented in the previous section, these images can be transformed to the computation of unrestricted images again.

A terminal case is reached whenever the problem has been reduced to a single component f_i. If f_i is not constant, then $\mathrm{Im}(f_i) = 1$. In the case $f_i = 1$ we have $\mathrm{Im}(f_i) = y_i$, and in the case $f_i = 0$ we have $\mathrm{Im}(f_i) = \overline{y_i}$. Furthermore, to improve the efficiency of output splitting the same techniques can be applied as in the case of input splitting.

Example 10.21. Let $f = (f_1, f_2, f_3)$ be the function from Example 10.19,

$$f_1 \; = \; x_1(x_2 + x_3), \qquad f_2 \; = \; x_2(x_1 + x_3), \qquad f_3 \; = \; x_3(x_1 + x_2).$$

The initial decomposition is

$$\text{Im}(f_1, f_2, f_3) = y_1 \cdot \text{Im}((\cdot, f_2, f_3), \text{on}(f_1)) + \overline{y_1} \cdot \text{Im}((\cdot, f_2, f_3), \text{on}(\overline{f_1}))$$
$$= y_1 \cdot \text{Im}((\cdot, f_2 \downarrow f_1, f_3 \downarrow f_1)) + \overline{y_1} \cdot \text{Im}((\cdot, f_2 \downarrow \overline{f_1}, f_3 \downarrow \overline{f_1})).$$

For $f_2^{(1)} = f_2 \downarrow f_1$ and $f_3^{(1)} = f_3 \downarrow f_1$ a computation yields $f_2^{(1)} = \overline{x_2}$ and $f_3^{(1)} = x_2 \overline{x_3} + \overline{x_2}$. Hence,

$$\text{Im}(\cdot, \cdot, f_3^{(1)} \downarrow f_2^{(1)}) = \text{Im}(\cdot, \cdot, x_3) = 1,$$
$$\text{Im}(\cdot, \cdot, f_3^{(1)} \downarrow \overline{f_2^{(1)}}) = \text{Im}(\cdot, \cdot, 1) = y_3,$$

and

$$\text{Im}((\cdot, f_2, f_3), \text{on}(f_1)) = y_2 \cdot \text{Im}(\cdot, \cdot, f_3^{(1)} \downarrow f_2^{(1)}) + \overline{y_2} \cdot \text{Im}(\cdot, \cdot, f_3^{(1)} \downarrow \overline{f_2^{(1)}})$$
$$= y_2 + \overline{y_2} y_3 = y_2 + y_3.$$

Due to $f_2 \downarrow \overline{f_1} = f_3 \downarrow \overline{f_1} = \overline{x_1} x_2 x_3$ during the computation of $\text{Im}((\cdot, f_2, f_3), \text{on}(\overline{f_1}))$, identical subproblems are recognized. This results in

$$\text{Im}((\cdot, f_2, f_3), \text{on}(\overline{f_1})) = y_2 y_3 + \overline{y_2}\ \overline{y_3}.$$

Finally, the total result can be determined by applying a disjunction:

$$\text{Im}(f_1, f_2, f_3) = y_1 \cdot \text{Im}((\cdot, f_2, f_3), \text{on}(f_1)) + \overline{y_1} \cdot \text{Im}((\cdot, f_2, f_3), \text{on}(\overline{f_1}))$$
$$= y_1(y_2 + y_3) + \overline{y_1}(y_2 y_3 + \overline{y_2}\ \overline{y_3})$$
$$= y_1 y_2 + y_1 y_3 + y_2 y_3 + \overline{y_1}\ \overline{y_2}\ \overline{y_3}.$$

Of course, the result coincides with that in Example 10.19. $\diamond$

10.4.3 The Transition Relation

The methods of input and output splitting are based on deducing the image directly from the functional description of the function $f = (f_1, \dots, f_m)$. Another approach starts by representing the transition behavior of the finite state machine M in terms of the characteristic function of a suitable relation. This relation describes all state pairs which are connected by an edge in the state diagram of M.

Definition 10.22. Let $M = (Q, I, O, \delta, \lambda, q_0)$ be a finite state machine with n state bits and p input bits. The **transition relation** $T_M : \mathbb{B}^{2n+p} \to \mathbb{B}^n$ of M is defined by

$$T_M(x, y, e) = T_M(x_1, \dots, x_n, y_1, \dots, y_n, e_1, \dots, e_p)$$
$$= \prod_{i=1}^{n} (y_i \equiv \delta_i(x, e)).$$

The variables $x_1, \dots, x_n$ are called **present-state variables** and the variables $y_1, \dots, y_n$ are called **next-state variables**.

Theorem 10.23. Let $M = (Q, I, O, \delta, \lambda, q_0)$ be a finite state machine with n state bits and p input bits, and let $C \subset \mathbb{B}^n$. The set C' of all states which can be reached from C in one step satisfies

$$\chi_{C'}(y) = \exists_{x_1, \dots, x_n} \exists_{e_1, \dots, e_p} (T_M(x, y, e) \cdot \chi_C(x)). \tag{10.4}$$

Proof. For a triple $(x, y, e) \in \mathbb{B}^{2n+p}$, we have $T_M(x, y, e) = 1$ if and only if in M the state x is the successor state of y for the input e. Consequently, a vector $y \in \mathbb{B}^n$ is contained in the set described by (10.4) if and only if there exists a vector $x \in C$, from which the state y can be reached for some input $e \in \mathbb{B}^p$.   □

The next-state function δ of a finite state machine M is a function $\mathbb{B}^{n+p} \to \mathbb{B}^n$. In order to be able to transfer the notation from the previous section with regard to image computation, we combine the inputs and the current states and consider again a function

$$f : \mathbb{B}^n \to \mathbb{B}^m$$

(where the symbol n now denotes the number of input and state bits, and m denotes the number of state bits). The transition relation $T_f(x, y)$ which corresponds to the function f is an element of the set $\mathbb{B}_{n+m}$. According to Theorem 10.23, this implies

$$\mathrm{Im}(f, C) = \exists_{x_1, \dots, x_n} (T_f(x, y) \cdot \chi_C(x)). \tag{10.5}$$

In the framework of this **image computation based on the transition relation**, we investigate, first, how far generalized cofactors can be used for improving efficiency as well.

Lemma 10.24. Let $h \in \mathbb{B}_{n+m}$, $g \in \mathbb{B}_n$, and $x = (x_1, \dots, x_n)$, $y = (y_1, \dots, y_m)$. Further, let ∇ be a positive generalized cofactor with

$$\exists_x (h(x, y) \nabla g(x)) \leq \exists_x (h(x, y) \cdot g(x)). \tag{10.6}$$

Then

$$\exists_x (h(x, y) \cdot g(x)) = \exists_x (h(x, y) \nabla g(x)).$$

Proof. According to Theorem 10.4 and Definition 10.6, the function $h\nabla g$ is a positive generalized cofactor of h with respect to g if and only if $(h\nabla g) \cdot g = h \cdot g$. This implies $(h \cdot g) \leq h\nabla g$, and, due to monotony of existential quantification,

$$\exists_x(h(x,y) \cdot g(x)) \leq \exists_x(h(x,y)\nabla g(x)).$$

It remains to prove the reverse inequation. However, this reverse inequation is exactly the statement (10.6) in the precondition. □

The question whether the constrain operator and the restrict operator satisfy property (10.6) can be answered positively. First we prove the statement for the constrain operator.

Theorem 10.25. *Let* $h \in \mathbb{B}_{n+m}$, $g \in \mathbb{B}_n$, *and* $x = (x_1,\dots,x_n)$, $y = (y_1,\dots,y_m)$. *Then*

$$\exists_x(h(x,y) \cdot g(x)) = \exists_x(h(x,y) \downarrow g(x)).$$

Proof. W.l.o.g. let $h \neq 0$, and let $x \in \mathbb{B}_n$, $y \in \mathbb{B}_m$. Then, by definition of the constrain operator, we have

$$h(x,y) \downarrow g(x) = h(s_1, s_2)$$

for some $s_1 \in \mathbb{B}^n, s_2 \in \mathbb{B}^m$ with $g(s_1) = 1$ and $||(x,y) - (s_1, s_2)||$ minimal. As the chosen s_1, s_2 only depend on g and not on h, the minimality of the distance implies $s_2 = y$. By using this intermediate result we can now prove condition (10.6) of Lemma 10.24.

Let $y \in \mathbb{B}^m$ which satisfies the statement $\exists_x(h(x,y) \downarrow g(x))$. Due to the above intermediate results there is an $s \in \mathbb{B}^n$ with

$$h(s,y) \cdot g(s) = 1.$$

This fact says nothing else but

$$\exists_s(h(s,y) \cdot g(s)) = 1$$

which proves condition (10.6). □

The next lemma and the next theorem show that the restrict operator can also be used for simplifying the image computation based on the transition relation.

Lemma 10.26. *Let* $h \in \mathbb{B}_{n+m}$, $g \in \mathbb{B}_n$ *and* $r \in \mathbb{B}^n$, $s \in \mathbb{B}^m$. *Further let* $\tilde{g} \in \mathbb{B}_n$ *such that*

$$h \Downarrow g = h \downarrow \tilde{g}.$$

If $\tilde{g}(r) > g(r)$, *then there exists some* $b \in \mathbb{B}^n$ *with*

$$\tilde{g}(b) = g(b) = 1 \quad and \quad h(r,s) = h(b,s).$$

Proof. As $\tilde{g}$ originates from g by existential quantification of some cofactors, the property $\tilde{g}(r) > g(r)$ implies that on the computation path of r a subset of the variables is quantified out. Let this subset be $J = \{x_{j_1}, \ldots, x_{j_p}\}$. h and $\tilde{g}$ do not depend on the variables in J. For h, this holds true because a quantification is only performed if the present cofactor of h does not depend on the present top variable in the order. For $\tilde{g}$ this holds true due to the quantifications performed.

Furthermore, there exists some input x with $x_i = r_i$ for all $x_i \notin J$ such that $g(x) = 1$. Otherwise, we would have $\tilde{g}(r) = g(r) = 0$ in contradiction to the precondition. As h does not depend on the variables J, there is some $b \in \mathbb{B}^n$ with $h(b, s) = h(r, s)$ and $g(b) = 1$.   □

Theorem 10.27. *Let* $h \in \mathbb{B}_{m+n}$, $g \in \mathbb{B}_n$, *and* $x = (x_1, \ldots, x_n)$, $y = (y_1, \ldots, y_m)$. *Then*

$$\exists_x (h(x, y) \cdot g(x)) = \exists_x (h(x, y) \Downarrow g(x)).$$

Proof. As in the proof of Theorem 10.25, we show that condition (10.6) of Lemma 10.24 is satisfied. Let $\tilde{g}$ be the function which satisfies $(h \Downarrow g)(x, y) = (h \downarrow \tilde{g})(x, y)$.

Let $y \in \mathbb{B}^m$ such that the statement $\exists_x (h(x, y) \downarrow \tilde{g}(x))$ holds true. Then there is some $s \in \mathbb{B}^n$ with

$$h(s, y) \cdot \tilde{g}(s) = 1.$$

If $g(s) = 1$ then

$$h(s, y) \cdot g(s) = 1 \cdot 1 = 1. \tag{10.7}$$

If $g(s) = 0$ then by Lemma 10.26 there is some $b \in \mathbb{B}^n$ with $\tilde{g}(b) = g(b) = 1$ and $h(b, y) = h(s, y) = 1$. Hence,

$$h(b, y) \cdot g(b) = 1 \cdot 1 = 1. \tag{10.8}$$

The two statements (10.7) and (10.8) immediately imply condition (10.6).   □

An advantage of the constrain operator is the property of distributivity which is proven in the next lemma.

Lemma 10.28. *Let* $h \in \mathbb{B}_m$, $f : \mathbb{B}^n \to \mathbb{B}^m$, $g \in \mathbb{B}_n$. *Then*

$$h(f(x)) \downarrow g(x) = h((f \downarrow g)(x)).$$

Proof. Let $x = (x_1, \ldots, x_n)$, and according to the definition of the constrain operator

$$h(f(x)) \downarrow g(x) = h(f(s))$$

for some $s \in \mathbb{B}^n$ with $g(s) = 1$. As the chosen s only depends on g and not on h or f, the same s satisfies

$$h((f \downarrow g)(x)) = h(f(s)).$$

$\square$

For the choice $h = x_1 \cdot x_2$ and $f = (f_1, f_2)$, Lemma 10.28 implies a distributive law with respect to conjunction and the constrain operator,

$$(f_1 \cdot f_2)(x) \downarrow g(x) = (f_1 \downarrow g)(x) \cdot (f_2 \downarrow g)(x).$$

Hence, in computations based on the transition relation the constrain operator can be applied to each factor in the transition relation,

$$\exists_x \left(\left(\prod_{i=1}^{m} (y_i \equiv f_i(x_1, \ldots, x_n)) \right) \downarrow g(x_1, \ldots, x_n) \right)$$

$$= \exists_x \left(\prod_{i=1}^{m} ((y_i \equiv f_i(x_1, \ldots, x_n)) \downarrow g(x_1, \ldots, x_n)) \right).$$

This property of distributivity does not hold true for the restrict operator.

10.4.4 Partitioning the Transition Relation

The main problem in image computation based on transition relations is to construct the OBDD of the transition relation according to Definition 10.22 or Equation (10.5). Even if the OBDDs of the individual functions f_i are small, the OBDD of the transition relation may become very large. In particular, this holds true if each of the functions f_i only depends on few variables, but the union of all these variables is large. Due to quantification the current-state variables are removed within the process of image computation, and the OBDD becomes substantially smaller during this process.

A strategy for avoiding these peak values of memory consumption is based on the following observation, which follows immediately from the definition of the Boolean existential quantifier.

Lemma 10.29. *Let $f \in \mathbb{B}_{n+m}$ be a function in the variables $x_1, \ldots, x_n$, $y_1, \ldots, y_m$, and let $g \in \mathbb{B}_{m+n-i+1}$ be a function in the variables $x_i, \ldots, x_n$, $y_1, \ldots, y_m$, $1 \leq i \leq n$. Then*

$$\exists_{x_1, \ldots, x_n} (f \cdot g) = \exists_{x_i, \ldots, x_n} \left(\exists_{x_1, \ldots, x_{i-1}} (f) \cdot g \right).$$

$\square$

The lemma suggests partitioning the transition relation into several blocks. Then image computation can be performed successively, where after each multiplication possibly some of the variables can be quantified early and hence removed.

In the following, let $f : \mathbb{B}^n \to \mathbb{B}^m$. Further, let $P_1, \dots, P_q$ be a partition of the set $\{1, \dots, m\}$ which reflects a partition of the components $f_1, \dots, f_m$ of f. The sets $P_1, \dots, P_q$ induce a decomposition of the transition relation into factors $T_1, \dots, T_q$ according to

$$T_i(x, y) = \prod_{i \in P_i} (y_i \equiv f_i(x)), \qquad 1 \le i \le q.$$

Obviously, the transition relation T_f of f can be expressed by

$$T_f(x, y) = \prod_{i=1}^{q} T_i(x, y).$$

Let $\pi = \pi(1), \dots, \pi(q)$ be an arrangement of the partition sets $\{P_1, \dots, P_q\}$. By V_i we denote the set of variables which are essential for at least one of functions in P_i. Further, let W_i denote the set

$$W_i = V_{\pi(i)} - \bigcup_{j=i+1}^{q} V_{\pi(j)}, \qquad 1 \le i \le q.$$

W_i contains all variables in $V_{\pi(i)}$ which are not essential for any of the functions in $P_{\pi(j)}$ with $j > i$.

By using these notations, the iterative computation of the set $\mathrm{Im}(f, C)$ by early quantification can be stated explicitly.

$$\chi_0(x, y) = \chi_C(x),$$
$$\chi_1(x, y) = \exists_{x_i : i \in W_1} (T_{\pi(1)}(x, y) \cdot \chi_0(x, y)),$$
$$\chi_2(x, y) = \exists_{x_i : i \in W_2} (T_{\pi(2)}(x, y) \cdot \chi_1(x, y)),$$
$$\vdots$$
$$\chi_q(x, y) = \exists_{x_i : i \in W_q} (T_{\pi(q)}(x, y) \cdot \chi_{q-1}(x, y)).$$

Finally, as in the set $\chi_q(x, y)$ all variables $x_1, \dots, x_n$ have been quantified, we have $\chi_q(x, y) = \chi_q(y) = \mathrm{Im}(\delta, C)$.

The choice of the partition $P_1, \dots, P_q$ and the order π in which the partition is processed have a big influence on the efficiency of image computation. As in the case of variable ordering, the computation of an optimal partitioning is not practical. Instead, there are some quite effective heuristics.

To determine the individual blocks T_i, one typically uses either a priori knowledge about the investigated circuit or a greedy strategy. In the greedy strategy, factors are added to a partial product until a given OBDD size is exceeded. Then a new block is introduced.

One of the most frequently used heuristics for ordering the partition sets goes back to Geist and Beer. Here, the idea is to make as many variables as possible ready for quantification in order to keep the OBDD size small during the whole iterative image computation. In the heuristic, $\pi(1), \pi(2), \ldots$ are determined successively. A variable x_i within this determination process is called *unique with respect to the partition* P_j if from the remaining partitions only P_j depends on x_i. The block with the largest number of unique variables is put at the beginning of the order. In case of a tie among several blocks we choose the block containing a maximal number of variables which do not occur in other blocks. Then the procedure is applied recursively to the remaining blocks until the order π is completely determined.

In practical use, the image computation based on a transition relation is superior to the two methods of input and output splitting. This fact mainly originates from two aspects. From the practical point of view, the optimization potential of building blocks and quantification can be realized quite effectively for the method based on a transition relation. From the theoretical point of view we have: the two splitting methods in principle perform an exponential number of decompositions, and the acceleration is solely based on implementation techniques like the use of a computed table. In the method based on a transition relation only a linear number of relatively well understood operations on OBDDs is necessary.

10.5 References

The analysis of sequential systems using symbolic OBDD techniques was established primarily by Coudert, Berthet, and Madre [CBM89] as well as by Burch, Clarke, Long, McMillan, and Dill [BCL$^+$94]. The constrain operator and the restrict operator also go back to Coudert, Berthet, and Madre. A survey article on implicit set representations can be found in [CM95].

The partitioning technique for image computation based on a transition relation is due to [BCL$^+$94]. The presented heuristic for the arrangement of the partitions was proposed by Geist and Beer [GB94].

Our presentation of reachability analysis follows the one in [Som96a].

11. Symbolic Model Checking

Corriger la fortune [Take control of destiny].
Gotthold Ephraim Lessing (1729–1781)

In the verification methods presented in Chapter 10, the model of finite state machine was the center of attention. Based on this, we now consider the more general verification concept of model checking. This concept is not strictly tied to the model of finite state machine, but is capable in addition of handling logic-based specifications.

We begin with an explanation of the term: the aim of **model checking** is to check whether an implementation satisfies a specification that is given by a logic formula. By considering a formalization of a specification within a formal logic, properties of a system can be described completely independent of concrete implementation details. Examples of those properties are invariants, liveness properties, or fairness properties.

The idea of designing algorithms for model checking that are based on OBDD data structures has been developed independently by several research groups. Due to the symbolic character of OBDD-based computations one also speaks of **symbolic model checking.**

In this chapter, we first discuss the temporal logic CTL which provides a suitable framework for model checking. The significance of this logic results from two facts. On the one hand, the formulas of this logic are suited quite well for specifying important properties of sequential systems. On the other hand, manipulations within this logic can be performed quite well in terms of OBDDs. Subsequent to this, in Section 11.2, we explain in detail how model checking can be realized efficiently by means of OBDD-based methods. This realization relies strongly on the fundamental techniques presented in Chapter 10. In Section 11.3, we describe some existing model checkers.

11.1 Computation Tree Logic

As sequential systems capture time-variant behavior, it is not possible to describe their properties completely in the framework of conventional propositional formulas. In a **temporal logic**, modal operators are additionally provided which can be used to express time-variant dependencies.

In particular, with regard to modeling time, two different types of temporal logics can be distinguished. In a **linear time temporal logic**, time is imagined as a factor which proceeds in a fixed direction in a linear manner, and which can be quantified by means of the real or the natural numbers. If the behavior of finite state machines were described in a linear time temporal logic, the operators could always refer to only a single sequence of states. In a **branching time temporal logic**, time is imagined to branch and proceed like a tree as shown in Fig. 11.1. Here, the branching points correspond to events which take place at measurable discrete points in time. The past of each event is uniquely determined, but the future is not. This corresponds exactly to the dynamic behavior of finite state machines. At each time the already traversed sequence of states is uniquely determined. As the future input values are not already determined, there are several alternatives for each step in the future.

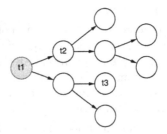

Figure 11.1. Illustration of the model of branching time. The time t_1 is reached before the times t_2 and t_3. But the times t_2 and t_3 are not in any temporal relation

Computation tree logic (CTL) is a temporal logic which is based on the model of branching time. The formulas of the logic describe properties of computation paths. Here, a **computation path** is an infinite sequence of states which is traversed during the computation.

In addition to the logic connections AND, OR, and NOT, the logic CTL has the following four operators which can be used to express temporal relations: the **next-time operator X**, the **global operator G**, the **future operator F**, and the **until operator U**. In a first stage, these operators are defined for a fixed computation path P and have the following meaning there:

Next-time operator X f: In the next step on the computation path P, the formula f holds true.

Global operator G f: f is valid globally in all states on the computation path P.

Future operator F f**:** f holds true eventually on the computation path P.

Until operator f **U** g**:** There is some state s on the computation path P in which g is valid, and f holds true in all states preceding s.

In general, more than one computation path starts in a given state. For this reason, each operator in CTL is preceded by a **path quantifier**. There are two path quantifiers: the *universal* path quantifier **A** and the *existential* path quantifier **E**.

Universal path quantifier A: The property holds true on *all* computation paths which start in the current state.

Existential path quantifier E: There is at least one computation path starting in the current state on which the property holds true.

As each of the temporal operators is preceded by a path quantifier, the truth value of a formula only depends on the present state and not on a specific computation path P.

Example 11.1. Consider the sequences of states in Fig. 11.2. If a state is labeled by a formula f, this tells us that in this state the formula f is valid. In the left representation the following statement for the distinguished state q_0 holds true: on each computation path starting in q_0 the formula f is eventually valid. As a consequence, the temporal formula **AF** f holds true in state q_0.

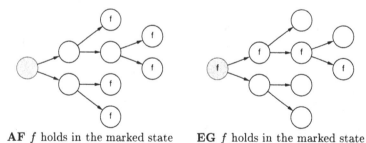

AF f holds in the marked state   EG f holds in the marked state

Figure 11.2. Path quantifiers

In the right representation there is a computation path starting in q_0 on which the formula f is globally valid. Hence, **EG** f holds true.   ◇

Indeed, already the operators **EX**, **E(U)** and **EG** suffice to describe the whole logic, as all other operators can be deduced from these three.

Lemma 11.2. *The following relations hold true:*

$$\mathbf{EF}\ g = \mathbf{E}(\mathrm{TRUE}\ \mathbf{U}\ g),$$
$$\mathbf{AX}\ f = \neg\mathbf{EX}\ (\neg f),$$
$$\mathbf{AG}\ f = \neg\mathbf{EF}\ (\neg f),$$
$$\mathbf{A}(f\ \mathbf{U}\ g) = \neg(\mathbf{E}(\neg g\ \mathbf{U}\ \neg f \wedge \neg g)\ \vee\ \mathbf{EG}\ \neg g), \tag{11.1}$$
$$\mathbf{AF}\ g = \mathbf{A}(\mathrm{TRUE}\ \mathbf{U}\ g), \tag{11.2}$$

where $\neg$ denotes the negation of a formula.

By using Property (11.1), the Property (11.2) can also be reduced to the three operators.

Proof. Exemplarily, we prove Property (11.1). The opposite of the statement $\mathbf{A}(f\ \mathbf{U}\ g)$ is: there is a computation path on which g never holds true, $(\mathbf{EG}\ \neg g)$, or there is a computation path on which in some state q neither f nor g is valid and in *no* state preceding q the formula g is valid. This is exactly Equation (11.1). □

Example 11.3. In this example, some typical CTL formulas are collected which are useful for proving the correctness of sequential systems.

$\mathbf{AG}\ (req \rightarrow \mathbf{AF}\ ack)$: Each request is eventually acknowledged.

$\mathbf{AG}\ \neg(f\ \mathrm{AND}\ g)$: This formula guarantees mutual exclusion. The properties f and g are never satisfied simultaneously.

$\mathbf{AG}\ (req \rightarrow \mathbf{A}\ (req\ \mathbf{U}\ ack))$: Each request is stored until an acknowledgment takes place.

$\mathbf{AG}\ \mathbf{EF}\ q_0$: At each time, there is a suitable sequence of states which allows a return to the initial state. Such a property serves to verify that the system can never *deadlock*, where deadlock means that the system stops working due to a waiting state which cannot be cleared. ◇

11.2 CTL Model Checking

Since the mid-1980s, decision procedures have been known which can be used to check whether a CTL formula is valid for a given sequential system. The particular importance of CTL among the temporal logics stems from the fact the decision procedures for CTL can be efficiently performed in the complexity theoretical sense.

In the original methods the state space was traversed enumeratively using depth first search. With the advance of OBDD technology, symbolic breadth first traversal as described in Chapter 10 could be employed in the framework of model checking. The basic strategy in model checking is based on computing state sets with certain temporal logical properties. We now explain an OBDD-based procedure which decides whether a given formula f is satisfied in a state s of a sequential system.

Let the transition relation of the system be denoted by T and represented by an OBDD: $T(s_0, s_1)$ is 1 if and only if s_1 is a successor state of s_0. The fact that we do not consider the input in the transition relation expresses that the formula under investigation should hold true independently of the input. Hence, formally, the transition relation T originates from Definition 10.22 by existential quantification over all primary inputs.

The algorithm is based on a function Check with the following specification:

Input: The transition relation T of a sequential system and a CTL formula f.

Output: The set of states of the sequential system which satisfy the formula f.

Of course, the representation of the state set is carried out in terms of OBDDs again.

The function Check is defined inductively over the structure of the CTL formulas. According to Lemma 11.2, only three CTL operators have to be considered. The inductive steps on the CTL formulas **EX** f, **E**$(f$ **U** $g)$, and **EG** f are computed by means of the auxiliary functions CheckEX, CheckEU, CheckEG:

$$\text{Check}(\textbf{EX } f, T) = \text{CheckEX}(\text{Check}(f), T),$$
$$\text{Check}(\textbf{E}(f \textbf{ U } g), T) = \text{CheckEU}(\text{Check}(f), \text{Check}(g), T),$$
$$\text{Check}(\textbf{EG } f, T) = \text{CheckEG}(\text{Check}(f), T).$$

Note that the arguments of the auxiliary procedures CheckEX, CheckEU, and CheckEG are propositional formulas, whereas as argument of Check more general CTL formulas are possible. If the argument of Check is a time-independent propositional formula, then the occurring Boolean operators are treated as in case of a symbolic simulation.

The formula **EX** f is true in a state x if and only if there exists a successor state of x which satisfies f. In Boolean notation this can be expressed by

$$\text{CheckEX}(f, T)(x) = \exists s_1 \left(T(x, s_1) \cdot \text{CheckEX}(f, T)(s_1) \right).$$

Of course, in the symbolic set representation in terms of OBDDs, all states with this property are processed simultaneously.

The definition of CheckEX has the same structure as image computation based on a transition relation, but there is one essential difference. In image computation, all states are determined which can be reached from a state

s_0 in one step, but here, all those states are computed starting in which one can enter the state s_1. Hence, an **inverse image computation**, also called a **backward step**, has to be performed. However, the techniques for efficient image computation from Chapter 10 can be adapted to inverse image computation.

The formula $\mathbf{E}(f \ \mathbf{U} \ g)$ means that on a computation path there exists a state x with the property g, and f is true in all states preceding x. Recursively, this fact can be expressed as follows:

- g is true in the current state, or
- f is true, and furthermore, there is a successor state in which $\mathbf{E}(f \ \mathbf{U} \ g)$ is satisfied.

This relation can be realized by means of a fixed point computation analogously to reachability analysis:

$$\text{CheckEU}_0(f, g, T)(x) = g(x),$$
$$\text{CheckEU}_{i+1}(f, g, T)(x) = \text{CheckEU}_i(f, g, T)(x) +$$
$$(f(x) \cdot \text{CheckEX}(\text{CheckEU}_i(f, g, T)(x))).$$

The number of states being described by CheckEU increases in a monotone way. As the state set is finite, there has to be a k with $\text{CheckEU}_{k+1} = \text{CheckEU}_k$. At this stage, the iteration can be stopped, and we have

$$\text{CheckEU}(f, g, T)(x) = \text{CheckEU}_k(f, g, T)(x).$$

The formula $\mathbf{EG} \ f$ is true in a state x if and only if there is a computation path starting in x on which f is always true. This means that f is true in the current state and that $\mathbf{EG} \ f$ is true in one of the successor states. Hence, the operator $\mathbf{EG}$ can also be expressed by means of a fixed point computation.

$$\text{CheckEG}_0(f, T)(x) = f(x),$$
$$\text{CheckEG}_{i+1}(f, T)(x) = \text{CheckEG}_i(f, T)(x) \cdot$$
$$(f(x) \cdot \text{CheckEX}(\text{CheckEG}_i(f, T)(x))).$$

In this sequence of monotone decreasing state sets, there has to be a k with $\text{CheckEG}_{k+1} = \text{CheckEG}_k$. At this stage, the iteration can be stopped, and we have

$$\text{CheckEG}(f, T)(x) = \text{CheckEG}_k(f, T)(x).$$

In this way, for a given sequential system and a formula f which has to be verified, all states are determined in which the property f is satisfied. If f is true in the initial state of the system, then the system satisfies the formula. By using representations in form of OBDDs, the occurring state sets can be represented compactly.

Example 11.4. Consider a finite state machine M with two state bits, one input bit and initial state q_0. Further, let the next-state function $\delta = (\delta_1, \delta_2)$ be given by

$$\delta_1((x_1, x_2), e) = \overline{x_1}\ \overline{x_2}\ e + x_1 \overline{x_2} e + x_1 x_2,$$
$$\delta_2((x_1, x_2), e) = x_1 \overline{x_2} e,$$

the initial state is $q_0 = 00$. The state diagram of M is depicted in Fig. 11.3. State 01 is not reachable from the initial state, and hence, it is not shown in the figure.

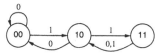

Figure 11.3. State diagram of a finite state machine M

Now the aim is to check whether from any reachable state the system can always return to the initial state q_0. The set S of the three states can be described in terms of the characteristic function χ_S,

$$\chi_S(x) = \chi_S(x_1, x_2) = x_1 + \overline{x_2}.$$

If the characteristic function (of the one-element set) of the state q_0 is denoted by χ_0, then we have

$$\chi_0(x) = \overline{x_1}\ \overline{x_2}\ .$$

Each state, from which the initial state q_0 is reachable, is characterized by **AF** $(\overline{x_1}\ \overline{x_2})$. We can return from each state to the initial state if the predicate

$$\chi_S(x) \rightarrow \mathbf{AF}\ (\chi_0(x)) \tag{11.3}$$

is a tautology.

According to Lemma 11.2 this statement can be written as

$$\chi_S(x) \rightarrow \mathbf{A}(\text{TRUE U } \chi_0(x))$$
$$\Longleftrightarrow \chi_S(x) \rightarrow \neg(\mathbf{E}(\overline{\chi_0(x)}\ \text{U (FALSE} \wedge \overline{\chi_0(x)})) \vee \mathbf{EG}\ \overline{\chi_0(x)})$$
$$\Longleftrightarrow \chi_S(x) \rightarrow \neg(\text{FALSE} \vee \mathbf{EG}\ \overline{\chi_0(x)})$$
$$\Longleftrightarrow \chi_S(x) \rightarrow \neg(\mathbf{EG}\ \overline{\chi_0(x)}). \tag{11.4}$$

For the existential global operator **EG** a fixed point procedure is at our disposal. The iteration for $(\mathbf{EG}\ \overline{\chi_0(x)})$ yields

$$\text{CheckEG}_0(\overline{\chi_0}, T)(x) = \overline{\chi_0(x)},$$
$$\text{CheckEG}_1(\overline{\chi_0}, T)(x) = \overline{\chi_0(x)} \cdot (\overline{\chi_0(x)} \cdot \text{CheckEX}(\text{CheckEG}_0(\overline{\chi_0}, T)(x), T))$$
$$= \overline{\chi_0(x)} \cdot \text{CheckEX}(\overline{\chi_0(x)}, T).$$

The computation of the inverse image results to

$$\text{CheckEX}(\overline{\chi_0(x)}, T) = \text{CheckEX}(x_1 + x_2, T)$$
$$= \overline{x_1}\,\overline{x_2} + x_1\overline{x_2}.$$

Hence, the iterative procedure further implies

$$\text{CheckEG}_1(\overline{\chi_0}, T)(x) = (x_1 + x_2) \cdot (\overline{x_1}\,\overline{x_2} + x_1\overline{x_2})$$
$$= x_1\overline{x_2},$$
$$\text{CheckEG}_2(\overline{\chi_0}, T)(x) = x_1\overline{x_2} \cdot (\overline{x_1}\,\overline{x_2}) \cdot \text{CheckEX}(\text{CheckEG}_1(\overline{\chi_0}, T)(x), T)$$
$$= 0.$$

Therefore $\text{CheckEG}_3(\overline{\chi_0}, T) = \text{CheckEG}_2(\overline{\chi_0}, T) = 0$, and the predicate $(\textbf{EG } \overline{\chi_0(x)})$ is a contradiction. Another equivalence transformation of (11.4) yields

$$\chi_S(x) \rightarrow \neg(\textbf{FALSE})$$
$$\Longleftrightarrow \chi_S(x) \rightarrow \textbf{TRUE}.$$

This shows that the predicate of interest (11.3) is a tautology. Therefore it has been formally proven that from each reachable state the system can always return to the initial state. By using the mentioned algorithms, all performed steps can be carried out completely automatically. ◇

11.3 Implementations

Based on the presented techniques, several OBDD-based model checkers have already been implemented and employed in industrial design cycles. A particular position is played by K. McMillan's symbolic model checker *SMV*, developed at Carnegie Mellon University, which has also been used in numerous other systems. The *VIS* system, primarily developed at the University of California at Berkeley and the University of Colorado at Boulder, unifies the mentioned verification techniques of finite state machines with techniques for synthesis of VLSI circuits. More recently, commercial systems have also become available, e.g., CVE (Circuit Verification Environment) by Siemens, or the system RuleBase by IBM which is built on top of SMV.

In the next two sections, the pioneering systems SMV and VIS will be presented in more detail.

11.3.1 The SMV System

SMV stands for **Symbolic Model Verifier**. This software system, developed by K. McMillan, has illustrated the capabilities and the power of OBDD-based verification tools with regard to several aspects. On the one hand, this system has served to verify pipeline-based arithmetic-logic units with more than 10^{100} (!) states. On the other hand, the cache coherence protocol of the IEEE (Institute of Electrical and Electronical Engineers) Futurebus+ Standard 896.1-1991 has been verified by using SMV. This protocol serves to ensure the consistency of local caches within a multiprocessor architecture. During the verification process several previously unknown bugs were discovered in the design of the protocol. Incidentally, this was the first time that an automatic verification tool found bugs in an IEEE standard.

By using the OBDD-based algorithms from Section 11.2, SMV allows to check sequential systems against specifications in the temporal logic CTL. The input language of SMV offers the possibility to define the systems which have to be analyzed on different abstraction levels.

To illustrate the input interface of SMV as well as its mode of operation, we consider the small demonstration program in Fig. 11.4. The input file contains both the description of the sequential system and its logic specification. The states of the finite state machine are represented by program variables. Here, not only binary but also multiple-valued variables are allowed. However, internally all variables are encoded in binary. In the example, the state of the system consists of two components, **request** and **state**. The component **request** can assume the two truth values **TRUE** or **FALSE**, the possible values for the component **state** are **ready** and **busy**.

In the example program the initial value of the variable **state** is set to **ready**. The successor value depends on the present state of the system. The value of a **case** expression is determined by the right side of the first line which

```
MODULE main
VAR
  request : boolean;
  state : {ready,busy};
ASSIGN
  init(state) := ready;
  next(state) := case
                   state = ready & request : busy;
                   1 : {ready,busy};
                 esac;
SPEC
  AG(request -> AF state = busy)
```

Figure 11.4. Input language of SMV

matches the condition on the left side. If state has the value ready and request is true, then the value busy is assigned to state. If this condition is not satisfied, then state will take on one of the two values ready or busy. This nondeterminism characterizes, e.g., the behavior as a reaction to different inputs to the system. Furthermore, we observe that the value of the variable request is never set explicitly in the program. Hence, this variable corresponds to a primary input variable.

The specification of this small system demands that each request eventually implies the value busy in the variable state. The SMV model checker verifies whether all possible initial states satisfy this specification.

11.3.2 The VIS System

VIS (Verification Interacting with Synthesis) is an OBDD-based software system which unifies verification, simulation and synthesis of finite state machines. It was developed primarily at the University of California at Berkeley and the University of Colorado at Boulder, and was made accessible to the public in 1996.

Figure 11.5 shows the architecture of the VIS system. There are four essential ingredients: the front end, the verification core, the synthesis core, and the underlying OBDD package.

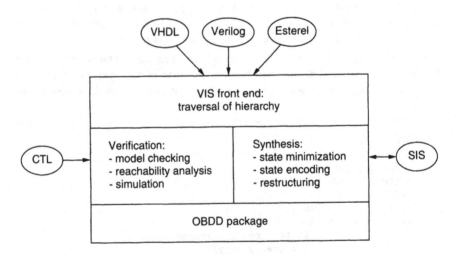

Figure 11.5. Architecture of the VIS system

The front end provides routines for transforming various well-known input formats into the internal hierarchical network description. In particular, also high-level languages like the widespread commercial language Verilog can be transformed into the internal format.

In order to represent the occurring switching functions, VIS employs OBDDs as internal data structure. All accesses to the level of Boolean manipulation are performed through uniform interfaces which correspond exactly to those operations and functions that have been discussed in the previous chapters. Due to the clear separation of the verification algorithms from the underlying level of Boolean manipulation it is possible to choose the OBDD package out of several available packages. In particular, the package of Long (see Section 7.2.2) and the CUDD package (see Section 7.2.3) are supported.

The verification core provides the presented algorithms for reachability analysis, equivalence checking, and model checking. In addition to the CTL-based model checking, fairness constraints, which cannot be expressed in CTL, can also be taken into account.

The synthesis core captures two aspects. On the one hand, it contains several new algorithms for synthesis of circuits which have been specifically developed for VIS. On the other hand, the synthesis core includes well-defined interfaces to the software package SIS (Sequential Interactive Synthesis). This software package, which is also based on OBDDs, was developed in Berkeley and provides numerous tools for synthesis and optimization of sequential circuits.

11.4 References

The significance of computation tree logic for model checking was recognized by Clarke, Emerson, and Sistla [CES86].

In the years 1989/90, several research groups simultaneously opened up the potential of OBDD-based set representations for model checking: Coudert, Madre, and Berthet, further Burch, Clarke, McMillan, and Dill, and as third group Bose and Fisher. A chronicle of these achievements can be found in [BCL+94]. A substantial contribution in the development of symbolic model checking is due to McMillan, the author of the OBDD-based model checker SMV [McM93].

The VIS verification system was developed by research groups in Berkeley and in Boulder under the direction of Brayton, Hachtel, Sangiovanni-Vincentelli, and Somenzi [BHS+96]. Descriptions of the commercial systems CVE and RuleBase can be found in [BLPV95] and [BBEL96].

12. Variants and Extensions of OBDDs

Wer nicht das Größere zum Großen fügt,
der möge nie sich seiner Ahnen rühmen.
[Those, who never join the greater
to the great one,
may never boast of their ancestors.]
August von Kotzebue (1761–1819): Oktavia

To further improve the efficiency of the data structure of OBDDs, several variants and extensions have been proposed. For the requirements in specific application fields, these refined models are better suited than the "classic" OBDDs. We would like to present some particularly interesting and important developments in this area, although the relevant research efforts have not been completed yet. The search for more compact representations of switching functions, which preserve the valuable properties of OBDDs, is still ongoing.

12.1 Relaxing the Ordering Restriction

Some important functions like multiplication of binary numbers or indirect storage access (e.g., the hidden weighted bit function in Definition 8.9) have provably exponential-size OBDDs with respect to any variable order. The proofs of these exponential lower bounds were given in Section 8.2. Now the aim is to investigate how far the restriction to a fixed linear order on the set of variables can be relaxed without weakening or even losing the good algorithmic properties of OBDDs.

If the ordering requirement as well as the read-once property of OBDDs are dropped completely, one immediately obtains general branching programs as introduced in Section 4.4.2. Of course, the size of an optimal branching program of a switching function f is less than or equal to the size of an optimal OBDD, but in many cases, the size may be much smaller, sometimes even exponentially smaller. However, this compactness often makes it hard or even impossible to perform the basic tasks of Boolean manipulation efficiently. For example, in Corollary 4.33 it was shown that the test whether two branching programs represent the same function is **co-NP**-complete.

The difficulty in performing the equivalence test for arbitrary branching programs is based on the fact that a variable may occur several times on a path. In order to avoid this problem, in Section 4.4.3 read-once branching programs were considered which test a variable at most once on each path. Indeed, in this scenario the equivalence test can be performed at least probabilistically in an efficient manner. However, for read-once branching programs a new difficulty arises. As proven in Theorem 4.38, performing binary operations for read-once branching programs is **NP**-hard. The reason for this difficulty can be seen in the fact that variables on different paths can occur in a different order.

The problem of performing a Boolean operation on two *arbitrary* read-once branching programs does not only emerge in this context. From Theorem 8.16 it is known that computing binary Boolean operations for two OBDDs with different variable orders is even an **NP**-hard problem. Only by adding the restriction that both OBDDs satisfy the same order does the problem become solvable in polynomial time. For this reason, J. Gergov and Ch. Meinel define the notion of a *complete type* for a read-once branching program which generalizes the notion of "order" of OBDDs in a sense that the following two properties are satisfied:

- With respect to a given complete type τ, read-once branching programs form a canonical representation of switching functions.
- For two read-once branching programs which are of the same type τ, binary operations can be performed efficiently.

As this concept yields a generalization of OBDDs, read-once branching programs are often denoted as **FBDDs (free binary decision diagrams)** in this context. The term "free" indicates that due to the read-once property it is true that whenever a variable is tested, this variable has not already been assigned a value before.

Definition 12.1. *(1) A **type** τ over the set of variables $\{x_1, \ldots, x_n\}$ is defined like an OBDD, but with the following two exceptions:*

1. *There is only one sink.*
2. *There does not need to exist a common variable order for the individual computation paths.*

As in the case of OBDDs, each variable may occur at most once on each path from the root to a sink.

*(2) A type τ is called **complete** if each variable occurs exactly once on each path from the root to a sink.*

(3) Let P be an FBDD. By merging the two sinks of P we obtain a type which is denoted by type(P).

First we remark that linear orders can be interpreted as quite simply structured complete types, see Fig. 12.1. Furthermore, the figure shows a more interesting example of a complete type with four variables.

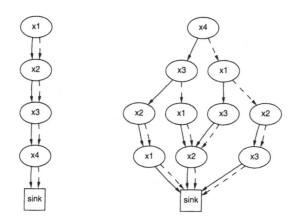

Figure 12.1. Two complete types

In Definition 6.8 the elimination and the merging rule were introduced for the reduction of OBDDs – the application of these rules does not modify the represented function itself. Indeed, the two reduction rules do not have any influence on the order in which the variables in the OBDD are read. For this reason, both rules can also be used in the context of FBDDs and their generalized orders described by types.

Definition 12.2. *We define two reduction rules on FBDDs and on types:*

Elimination rule: *If the 1-edge and the 0-edge of a node v point to the same node u, then eliminate v, and redirect all incoming edges of v to u.*

Merging rule: *If the internal nodes u and v are labeled by the same variable, their 1-edges lead to the same node and their 0-edges lead to the same node, then eliminate one of the two edges u, v, and redirect all incoming edges of this node to the remaining one.*

Definition 12.3. *(1) An FBDD P is called* **reduced** *if neither of the two reduction rules can be applied.*

(2) A type τ or an FBDD P are called **algebraically reduced** *if the merging rule cannot be applied to τ or P, respectively.*

The following statement can now be proven similarly to the uniqueness theorem of OBDDs.

Theorem 12.4. *The reduced FBDD which results from an FBDD P by com-plete application of the two reduction rules is determined uniquely. In the same way, the algebraically reduced FBDD and the algebraically reduced type which originate from complete application of the merging rules are uniquely determined.* ☐

In order to identify classes of FBDDs which test the variables in a similar sequence, it suffices to compare the types of different FBDDs. We define the notion of a subtype which serves to describe the consistency of types in different FBDDs.

Definition 12.5. *Let τ be a type. A type τ' is called a **subtype** of τ, τ' ≤ τ, if τ = τ' or if τ' can be constructed from τ by applications of the merging and the elimination rules.*

With respect to the operator ≤ the set of all types constitutes a partial order. Figure 12.2 shows a type τ and a subtype τ'.

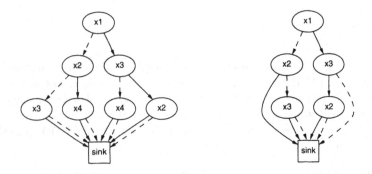

Figure 12.2. A type τ and a subtype τ', τ' ≤ τ

Definition 12.6. *Let τ be a complete type. An FBDD P **is of type** τ if there exists a type τ' with type(P) ≤ τ' such that τ and τ' coincide after a complete algebraic reduction.*

Due to this definition it is reasonable only to work with complete types which are already algebraically reduced.

One can see that in the sense of the definition an FBDD can belong to several types. However, if an FBDD P is of type τ, then the *reduced* FBDD which originates from P is also of type τ.

Equipped with these notational tools it can now be proven that reduced FBDDs with respect to a given complete type constitute a canonical repre-sentation of switching functions.

We begin with an observation: for representing any input variable x_i there is a decision diagram which consists of exactly one node labeled by x_i, and whose 1-edge and 0-edge lead to the 1-sink and 0-sink, respectively. Let us call this decision diagram (which is even ordered) the **standard representation** of x_i.

Lemma 12.7. *Let τ be a complete type over the set of variables $\{x_1, \ldots, x_n\}$, and let $x_i \in \{x_1, \ldots, x_n\}$. The standard representation P_{x_i} of x_i is of type τ.*

Proof. As τ is a complete type, the variable x_i appears on each path from the root to the sink. By means of the elimination rule, starting in the sink of τ, we can eliminate all successors of the node labeled by x_i. After iterated application of the merging rule we obtain a type τ' such that a single node labeled by x_i is the only predecessor of the sink. Finally, by iterated application of the elimination rule, this time starting in the node labeled by x_i, we can successively construct from τ' the type $type(P_{x_i})$. $\square$

Theorem 12.8. *Let τ be a complete and algebraically reduced type over the set of variables $\{x_1, \ldots, x_n\}$, and let $f \in \mathbb{B}_n$. Then there is exactly one reduced FBDD of type τ which represents the function f.*

Proof. First, it is clear that there exists a complete binary decision tree T which represents f and which is of type τ. Due to the completeness of τ and T, the tree T is uniquely determined. The reduced FBDD P of T is also of type τ, and according to Theorem 12.4 it is uniquely determined.

Now let P' be an arbitrary FBDD of type τ which represents f. The reduced FBDD P'' originating from P is also of type τ and represents f. Hence, Theorem 12.4 implies $P'' = P$. We can conclude that there is only one reduced FBDD of type τ which represents f. This FBDD can, e.g., be constructed by complete reduction of T. $\square$

This uniqueness statement is the reason why many OBDD-based algorithms, such as algorithms for performing binary operations, can be directly transferred to type-based FBDDs. We summarize the corresponding statements which result from generalizing the statements of OBDDs.

Theorem 12.9. *Let $f, f_1, f_2 \in \mathbb{B}_n$, and let τ be a complete, algebraically reduced type on the set of variables $\{x_1, \ldots, x_n\}$.*

Universality of FBDD representation: *Any switching function $f \in \mathbb{B}_n$ can be represented by a (reduced) FBDD of type τ.*

Canonicity of FBDD representation: *Any switching function $f \in \mathbb{B}_n$ has exactly one representation as a reduced FBDD of type τ.*

Efficient synthesis of FBDDs: *Let P_1, P_2 be two FBDD representations of f_1, f_2 which are of type τ. Then for each binary operation $*$ the reduced FBDD representation of $f = f_1 * f_2$ can be computed in time $\mathcal{O}(size(\tau) \cdot size(P_1) \cdot size(P_2))$.*

Efficient equivalence test: *Let P_1 and P_2 be two FBDD representations of f_1, f_2 which are of type τ. Equivalence of P_1 and P_2 can be decided in linear time $\mathcal{O}(size(P_1) + size(P_2))$. In case of a strongly canonical representation only constant time is needed.* $\qquad\square$

Compared to the more restricted class of OBDDs, the time complexity for computing binary operations increases by the additional factor $size(\tau)$. However, often binary operations in the case of FBDDs can be computed not only in cubic but even in quadratic time. For example, this holds true if the width of the underlying type is bounded by a constant.

When comparing OBDD and FBDDs, one should always keep in mind that OBDDs can be considered as special cases of FBDDs. For this reason, (optimal) FBDD representations are never larger than (optimal) OBDD representations, but FBDD representations may be substantially smaller. For the practical work with the representations, one should take into account that often the linear order of OBDDs can be handled more easily. This applies to the construction of heuristics, dynamic optimization algorithms, or more complex manipulation algorithms such as quantification. Hence, before tackling a certain practical application, it should be checked whether the additional optimization potential of FBDDs is in good relation to the additional expenses in the algorithmic handling.

We would like to present a concrete example which illustrates the additional optimization potential of FBDDs for a concrete function. For this purpose, we consider again the hidden weighted bit function $HWB(x)$, from Definition 8.9 which describes an indirect storage access. According to Theorem 8.10 it is known that with respect to any variable order the OBDD representation of $HWB(x)$ grows exponentially in the input length n.

In contrast to this, it is possible to represent the hidden weighted bit function by an FBDD of merely *polynomial* size. The basic idea of the construction is a recursive decomposition of the function $HWB(x) = HWB(x_1, \dots, x_n)$. To simplify notation, let $x_{i:j}$ denote the subsequence $x_i, \dots, x_j$ of the input vector $(x_1, \dots, x_n)$. We define the two functions $H_{i,j}, G_{i,j}$:

$$H_{i,j}(x_{i:j}) = \begin{cases} x_{i+wt(x_{i:j})-1} & \text{if } wt(x_{i:j}) > 0, \\ 0 & \text{otherwise,} \end{cases}$$

$$G_{i,j}(x_{i:j}) = \begin{cases} x_{i+wt(x_{i:j})} & \text{if } wt(x_{i:j}) < j - i + 1, \\ 1 & \text{otherwise.} \end{cases}$$

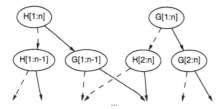

Figure 12.3. Construction of an FBDD of the hidden weighted bit function

The definition of $H_{i,j}$ immediately implies

$$HWB(x_1, \ldots, x_n) = H_{1,n}(x_{1:n}).$$

Now, a recursive construction rule for the functions $H_{i,j}$ and $G_{i,j}$, $i > j$, can be stated. This rule is based on Shannon's decomposition.

$$H_{i,j}(x_{i:j}) = \overline{x_j} H_{i,j-1}(x_{i:j-1}) + x_j G_{i,j-1}(x_{i:j-1}), \qquad (12.1)$$
$$G_{i,j}(x_{i:j}) = \overline{x_i} H_{i+1,j}(x_{i+1:j}) + x_i G_{i+1,j}(x_{i+1:j}). \qquad (12.2)$$

The terminal cases are

$$H_{i,i}(x_{i:i}) = G_{i,i}(x_{i:i}) = 1.$$

The idea of the recursive definition of $H_{i,j}$ and $G_{i,j}$ is to reduce the computation of the hidden weighted bit function to the computation of subproblems for which partial information is already known. The simultaneous computation of these subproblems with respect to different constraints finally allows the construction of a polynomial-size FBDD.

The recursive construction of an FBDD of $HWB(x_1, \ldots, x_n)$ is illustrated in Fig. 12.3. Simultaneously, the functions $H_{i,j}$ and $G_{i,j}$ are computed. For each function $H_{i,j}(x_{i:j})$ or $G_{i,j}(x_{i:j})$, which originates by k decompositions according to (12.1) or (12.2) from $H_{1,n}(x_1, \ldots, x_n) = HWB(x_1, \ldots, x_n)$, we have $j - i = n - k - 1$, $0 \le k \le n - 1$. Consequently, in the $(n - k)$-th level of the FBDDs there are only nodes which refer to the subfunctions $H_{i,i+k}(x_{i:i+k})$, $G_{i,i+k}(x_{i:i+k})$, $0 \le k \le n - 1$. For each k this amounts to at most $2n$ nodes. Furthermore, we observe that on each path from the root to the sink each variable is read at most once. Hence, we can conclude:

Theorem 12.10. *There is an FBDD representation of the hidden weighted bit function $HWB(x) = HWB(x_1, \ldots, x_n)$, whose number of nodes is bounded by the polynomial $2n^2 + 2$.* $\qquad \square$

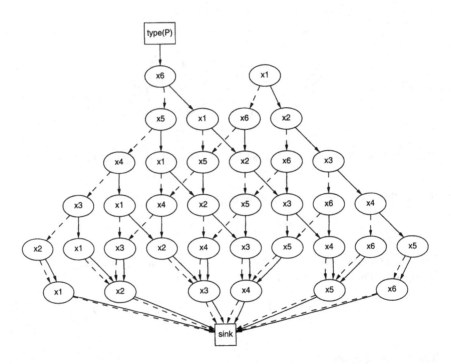

Figure 12.4. Complete type of the FBDD P for the hidden weighted bit function in case $n = 6$

Figure 12.4 shows the underlying complete type of the FBDD representation for the case $n = 6$. This type is tailored quite specifically for the hidden weighted bit function. In general, it is surely not practical to construct for each concrete application instance the best possible type. However, as analogously to the variable order of OBDDs the size of FBDDs depends very strongly on the chosen type, heuristic methods for automatically generating FBDD types are required. Bern, Gergov, Meinel, and Slobodová have presented such heuristics. They are based on ideas similar to those on which the heuristics for constructing good variable orders are based.

However, in case of FBDD heuristics an additional idea can be brought into play. If for example the top variable x_i of the type τ under construction has already been determined, then the construction of τ can be *independently* continued for the two situations $x_i = 0$ and $x_i = 1$. Finally, the type τ is obtained by a simply plugging together the two partial types which have been constructed.

12.2 Alternative Decomposition Types

If f denotes the function being represented by a node with label x_i in an OBDD, then Shannon's decomposition holds:

$$f = x_i f_0 + \overline{x_i} f_1$$

with the cofactors $f_0 = f(x_1, \ldots, x_{i-1}, 0, x_{i+1}, \ldots, x_n)$ and $f_1 = f(x_1, \ldots, x_{i-1}, 1, x_{i+1}, \ldots, x_n)$.

Already in Corollary 3.24 we presented some different methods for decomposing a function with respect to a given variable. We will now introduce further decomposition types, namely the *Reed-Muller* or *Davio decompositions*.

Theorem 12.11. *Let $f \in \mathbb{B}_n$ be a switching functions in n variables. For the functions f_0, f_1, f_2 defined by*

$$f_0(x) = f(x_1, \ldots, x_{n-1}, 0),$$
$$f_1(x) = f(x_1, \ldots, x_{n-1}, 1),$$
$$f_2(x) = f_0(x) \oplus f_1(x)$$

the following two decompositions hold true:

Reed-Muller decomposition *or* **positive Davio decomposition:**

$$f = f_0 \oplus x_n f_2.$$

Negative Davio decomposition:

$$f = f_1 \oplus \overline{x_n} f_2.$$

Of course, this decomposition holds in analogous manner for each variable x_i, $1 \leq i \leq n$.

Proof. Due to Shannon's decomposition with respect to the EX-OR operation $\oplus$ from Corollary 3.24, we have

$$
\begin{aligned}
f(x) &= \overline{x_n} f_0(x) \oplus x_n f_1(x) && (\overline{x_n} = 1 \oplus x_n) \\
&= f_0(x) \oplus (x_n f_0(x) \oplus x_n f_1(x)) \\
&= f_0(x) \oplus x_n f_2(x).
\end{aligned}
$$

Analogously,

$$
\begin{aligned}
f(x) &= x_n f_1(x) \oplus \overline{x_n} f_0(x) && (x_n = 1 \oplus \overline{x_n}) \\
&= f_1(x) \oplus (\overline{x_n} f_1(x) \oplus \overline{x_n} f_0(x)) \\
&= f_1(x) \oplus \overline{x_n} f_2(x).
\end{aligned}
$$

$\square$

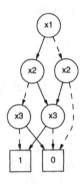

Figure 12.5. Example of an OFDD

Ordered functional decision diagrams introduced by U. Kebschull, E. Schubert, and W. Rosenstiel are not based on Shannon's decomposition like OBDDs, but instead are based on the Reed-Muller decomposition.

Definition 12.12. An ordered functional decision diagram, OFDD, *is defined like an OBDD with one difference: the function f_v which is computed in a node v of the graph is now defined by the following inductive rules:*

1. *If v is a sink with label 1 (0), then $f_v = 1$ ($f_v = 0$).*
2. *If v is a node with label x_i, whose 1- and 0-successor nodes represent the functions h and g, respectively, then*

$$f_v = g \oplus x_i h.$$

Example 12.13. If the graph in Fig. 12.5 is interpreted as an OBDD, then the function $f^{OBDD}(x) = x_1 x_2 + x_1 \overline{x_2} \ \overline{x_3}$ is represented. But if the graph is interpreted as an OFDD, the two nodes with label x_3 represent the functions

$$f_{x_3,1}(x) = 1 \oplus x_3 \cdot 0 = 1, \qquad f_{x_3,2}(x) = 0 \oplus x_3 \cdot 1 = x_3,$$

the two nodes with label x_2 represent the functions

$$f_{x_2,1}(x) = 1 \oplus x_2 x_3, \qquad f_{x_2,2}(x) = 0 \oplus x_2 x_3 = x_2 x_3,$$

and the root node represents the function

$$f^{OFDD}(x) = x_2 x_3 \oplus x_1 (1 \oplus x_2 x_3) = x_2 x_3 \oplus x_1 \oplus x_1 x_2 x_3.$$

The represented functions $f^{OBDD}(x)$ and $f^{OFDD}(x)$ are totally different. For example, we have $0 = f^{OBDD}(1,0,1) \neq f^{OFDD}(1,0,1) = 1$. ◇

The elimination rule for OBDDs allows to remove nodes v with identical 1- and 0-successors without modifying the represented function. The reduction is possible if and only if the function being represented in v does not depend on the label variable of v. In case of OFDDs this reduction rule cannot be applied in this form, as in the case of two identical successors the function represented in v may depend on the label variable nevertheless. Instead, those nodes are redundant whose 1-edge points to the 0-sink, see Fig. 12.6.

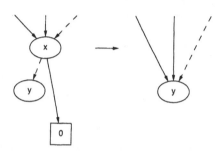

Figure 12.6. Elimination rule for OFDDs

Definition 12.14. *We define two reduction rules on OFDDs:*

Elimination rule: *If the 1-edge of a node v points to the 0-sink, then eliminate v, and redirect all incoming edges of v to the 0-successor of v.*

Merging rule: *If the internal nodes u and v are labeled by the same variable, their 1-edges lead to the same node and their 0-edges lead to the same node, then eliminate one of the two edges u, v, and redirect all incoming edges of this node to the remaining one.*

Definition 12.15. *An OFDD P is called* **reduced** *if neither of the two reduction rules can be applied.*

Now in the same way as for OBDDs, a uniqueness statement can be deduced for OFDDs.

Corollary 12.16. *For all variable orders π we have: the reduced OFDD of a switching function f with respect to the order π is uniquely determined (up to isomorphisms).*   □

OFDDs do not only possess the property of canonicity, but they have also many other properties in common with OBDDs. However, there are some important differences. The first one refers to the application of binary operations. According to Theorem 6.16 all binary operations on OBDDs can be computed in polynomial time and space. In case of OFDDs this also holds for the EX-OR operation $\oplus$. However, the computation of a conjunction or a disjunction of two OFDDs can provably lead to an exponential increase in size.

Function evaluation for a given OBDD P and a given input vector $a = (a_1, \ldots, a_n) \in \mathbb{B}^n$ can be achieved quite easily: in each node with label x_i one follows the edge with label a_i until a sink is reached. The value of the sink gives the function value of P on the input a. In case of OFDDs it is not sufficient for evaluating an input to traverse only a single path from the root to a sink. Instead, for each node v *both* subgraphs have to be evaluated, and the exclusive-or (EX-OR) of both values has to be computed. However, by traversing the nodes level by level from the sinks to the root, an evaluation of the function is possible in asymptotically linear time in the number of nodes.

For some classes of functions, OFDDs are exponentially more compact than OBDDs, for other classes exactly the contrary is true. For this reason, Drechsler, Sarabi, Theobald, et al. have proposed a way to unite the advantages of both classes. They have shown that the different decomposition types can be combined within the same subgraph while preserving good algorithmic properties. In this hybrid representation type called **ordered Kronecker functional decision diagrams (OKFDDs)** each variable x_i is assigned a **decomposition type** $d_i \in \{S, pD, nD\}$ with the following meaning:

- S: Shannon decomposition,
- pD: positive Davio decomposition,
- nD: negative Davio decomposition.

In each node with label x_i the decomposition defined by d_i is carried out.

Example 12.17. Figure 12.7 shows the graph from Fig. 12.5 again, but this time additionally, each variable x_i is assigned a decomposition type S, pD or nD. In the OKFDD which is defined by this graph the two nodes with label x_3 represent the functions

$$f_{x_3,1}(x) = \overline{x_3}, \qquad f_{x_3,2}(x) = x_3,$$

the two nodes with label x_2 represent the functions

$$f_{x_2,1}(x) = \overline{x_3} \oplus \overline{x_2} x_3, \qquad f_{x_2,2}(x) = 0 \oplus \overline{x_2} x_3 = \overline{x_2} x_3,$$

and the root node represents the function

$$f^{OKFDD}(x) = \overline{x_2} x_3 \oplus x_1 (\overline{x_3} \oplus \overline{x_2} x_3).$$

$\diamond$

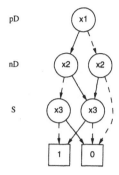

Figure 12.7. An OKFDD with the decomposition types $d_1 = pD$, $d_2 = nD$, $d_3 = S$

The classes of OFDDs and OKFDDs are particularly suited as data structure in connection with the Reed-Muller expansions described in Section 4.2.

Definition 12.18. *A path from the root of an OFDD to the 1-sink is denoted as* **1-path**.

The defining equation of the positive Davio decomposition immediately implies that an OFDD is basically a Reed-Muller expansion which has been written in a compact graph-based form. Hence, we can deduce the following theorem:

Theorem 12.19. *The number of 1-paths in an OFDD of the switching function* $f \in \mathbb{B}_n$ *is equal to the number of monomials in the Reed-Muller expansion of* f. □

Additionally, the data structure of OKFDDs allows us to solve problems in connection with **fixed-polarity Reed-Muller expansions** efficiently. Here, for each variable a fixed polarity (negated or unnegated) is chosen. In the discussed "classic" Reed-Muller expansions each variable occurs in unnegated form.

12.3 Zero-Suppressed BDDs

Combinatorial problems can be reduced to the task of manipulating switching functions, too. Here, sets of element combinations often have to be represented.

A **combination** of n elements in the bit vector representation is an n-bit vector $(x_1, \dots, x_n) \in \mathbb{B}^n$, where the i-th bit reports whether or not the i-th element is contained in the combination.

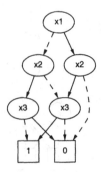

Figure 12.8. An OBDD and a ZDD of the set of combinations $\{(0,0,1),$ $(0,1,0),(1,0,0),(1,0,1)\}$

Hence, a set of combinations can be represented by a switching function $f :$ $\mathbb{B}^n \to \mathbb{B}$. A combination given by the input vector $(a_1, \ldots, a_n)$ is contained in the set if and only if $f(a_1, \ldots, a_n) = 1$.

Example 12.20. We consider the three elements $\{1, 2, 3\}$ which are represented by the variables x_1, x_2, x_3, respectively. Let t_1 be the combination in which the first and the second element are not contained, but the third element is, i.e., $t_1 = (0, 0, 1)$. In the same way, let us define the combinations $t_2 = (0, 1, 0)$, $t_3 = (1, 0, 0)$, $t_4 = (1, 0, 1)$. The corresponding Boolean function which represents the set $\{t_1, t_2, t_3, t_4\}$ is

$$f(x) = \overline{x_1}\,\overline{x_2}x_3 + \overline{x_1}x_2\overline{x_3} + x_1\overline{x_2}\,\overline{x_3} + x_1\overline{x_2}x_3.$$

An OBDD of f is depicted in Fig. 12.8.   ◇

In many applications with combinatorial background the bit vectors to be represented are sparse with regard to two aspects:

- the set to be represented contains only a small fraction of the 2^n possible bit vectors, and
- each bit vector contains many zeroes.

This is true since the underlying basis set of all elements may be quite large, while the number of really relevant combinations is often very small in comparison to this large set. Furthermore, often each of the relevant combinations consists of only a few elements. Based on these observations, S. Minato has introduced a BDD-based representation type which exploits both types of sparsity. **Zero-suppressed BDDs (ZDDs)**, sometimes also called **combinational sets**, are defined like OBDDs, but now the elimination rule is specifically adjusted in order to exploit sparsity.

Definition 12.21. *We define two reduction rules on ZDDs:*

Elimination rule: *If the 1-edge of a node v points to the 0-sink, then eliminate v, and redirect all incoming edges of v to the 0-successor of v.*

Merging rule: *If the internal nodes u and v are labeled by the same variable, their 1-edges lead to the same node and their 0-edges lead to the same node, then eliminate one of the two edges u, v, and redirect all incoming edges of this node to the remaining one.*

Hence, the elimination rule of ZDDs is of the same form as the elimination rule of OFDDs, see Fig. 12.6.

Example 12.22. If interpreted as OBDD, the graph in Example 12.20 (Fig. 12.8) represents the set of combinations

$$\{(0,0,1),(0,1,0),(1,0,0),(1,0,1)\}.$$

In order to interpret the graph as ZDD, we imagine adding a node with label x_3 on the edge from the node with label x_2 to the 0-sink – the 1-edge of the new node shall lead to the 0-sink. This node has been reduced due to the elimination rule of ZDDs. It can be easily seen that in this case the ZDD describes the same set of combinations as the OBDD.

We now extend the basis set formally by a fourth element which is represented by the bit x_4. If one now considers the same set as before, then all combinations have to be extended by a zero component, as x_4 occurs in none of these combinations:

$$\{(0,0,1,0),(0,1,0,0),(1,0,0,0),(1,0,1,0)\}.$$

The OBDD of the resulting Boolean function is depicted in Fig. 12.9. Starting from the OBDD in Fig. 12.8, it has to be additionally checked that for a combination in the set the property $x_4 = 0$ is satisfied.

Now we turn to the question of what the ZDD of the new set of combinations looks like. One easily checks that it is identical with the ZDD in Fig. 12.8 ! The formal extension by the fourth element to the basis set does not become noticeable at all, because the resulting nodes in the ZDD are suppressed by means of the elimination rule.

The example shows not only that ZDDs are independent of possible extensions to the basic set but also that ZDDs are particularly compact if only a small part of the basic set has to be represented. $\diamond$

Analogously to OBDDs, the class of ZDDs constitutes a canonical representation of switching functions. However, we would like to point out some important differences to OBDDs. In case of OBDDs a function can be reconstructed solely from a given graph. In contrast to this, in the case of ZDDs

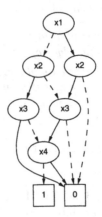

Figure 12.9. An OBDD of the set of combinations $\{(0,0,1,0),(0,1,0,0),$ $(1,0,0,0),(1,0,1,0)\}$

additional information about the underlying basis set of variables is required. Due to the elimination rule of ZDDs a graph which only consists of the 1-sink can represent the functions 1, $\overline{x_1}$, $\overline{x_1}\,\overline{x_2}$, $\overline{x_1}\,\overline{x_2}\,\overline{x_3}$, and so on. Which of these functions is represented in a certain case depends on the given basis set of variables.

In case of reduced ZDDs it is possible that the 1-edge and the 0-edge of a node point to the same successor. Those nodes can be reduced only in the case of OBDDs. The elimination rule defines for both types of graphs how the case of non-occurrence of a variable x_i has to be interpreted. This situation means

- in case of OBDDs: the represented function is independent of x_i;
- in case of ZDDs: $x_i = 1$ immediately implies the function value 0.

So far, the data structure of zero-suppressed BDDs has primarily been used to solve problems in two- and multiple-level logic minimization. An example from a quite different area illustrates the fundamental importance of ZDDs: M. Löbbing and I. Wegener report successful ZDD experiments for solving difficult combinatorial problems which occur in the analysis of knight moves on a chess board.

The following theorem says that when using ZDDs instead of OBDDs the size of the resulting graphs can decrease by at most a factor of n. However, in practical applications this seemingly small reduction can be large enough to decide the success of a computation.

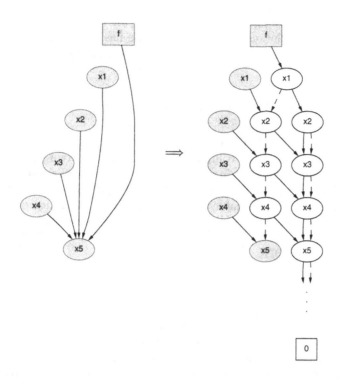

Figure 12.10. Construction of an OBDD from a ZDD, showing how to replace all possible edge types into a node with label x_5

Theorem 12.23. *Let $f \in \mathbb{B}_n$ and π be a variable order on $\{x_1, \dots, x_n\}$. The reduced OBDD P_O and the reduced ZDD P_Z of f with respect to π satisfy*

$$size(P_O) \leq n \cdot size(P_Z + 2).$$

Proof. Starting from the reduced ZDD P_Z we add those nodes again which do not appear due to the elimination rule. This addition of nodes does not modify the represented function.

W.l.o.g. we can assume the order to be the natural variable order $x_1, \dots, x_n$. To simplify notation, we further assume that the two sinks are labeled by an auxiliary variable x_{n+1}. For each node with label x_j, $1 \leq j \leq n+1$, we add $j-1$ nodes $v_1, \dots, v_{j-1}$ with labels $x_1, \dots, x_{j-1}$ to the ZDD, see Fig. 12.10. The 0-edge of the node v_k leads to the node v_{k+1}, where $v_j := v$. The 1-edge of the node v_k leads to that specific node v_{k+1} which has been constructed for the 0-sink. Finally, each edge of a node w with label x_i, $i < j-1$, to the node v is replaced by an edge to the node v_{i+1}. Here, each edge which points to the root of the ZDD is considered as an edge which stems from a node that is labeled by an auxiliary variable x_0.

In this way, the outgoing edges of each node with label x_i point to a node with label x_{i+1}. In the resulting graph no node has been eliminated due to the reduction rule. Hence, the graph is also an OBDD of the function f. □

12.4 Multiple-Valued Functions

Based on the advances that have been achieved by using OBDD data structures, the search for more compact, but still efficient representations of switching functions is continuing. In addition to this, there are several concepts for transferring the efficient OBDD-based algorithms also to applications involving non-binary values of functions. Here, the range of integers plays a prominent role. In the variants of OBDDs defined below, the input variables are kept binary in order to be able to establish a similar branching structure like in OBDDs.

12.4.1 Additional Sinks

A natural method for representing multiple-valued functions by decision graphs consists in adding further sinks. These sinks can be labeled by arbitrary values. If for example a function $f : \mathbb{B}^n \to \{0, \dots, n\}$ is to be represented, then the $n+1$ sinks carry the labels 0 through n. This representation type is denoted as **multi-terminal BDD (MTBDD)** or **algebraic decision diagram (ADD)**.

The evaluation of an MTBDD for a given assignment to the variables proceeds analogously to the evaluation in an OBDD: one pursues that unique path from the root to a sink which is determined by the input. The value of the reached sink givens the desired function value.

Example 12.24. Figure 12.11 illustrates an MTBDD representation of the function $x_0 + 2x_1 + 4x_2$ with respect to the variable order x_2, x_1, x_0. This function interprets the three input bits x_2, x_1, x_0 as binary representation and produces as result the corresponding natural number. For example, the input $x_2 = 1, x_1 = 0, x_0 = 1$ yields the value 5.   ◇

The example illustrates that MTBDDs are quite inefficient when representing functions with a large range. For example, if all those numbers are possible as function values whose binary representation has length at most n, then there exist 2^n different function values. Hence, each MTBDD representation of such functions contains an exponential number of terminal nodes. Often, however, the number of possible function values is much more restricted, and MTBDD representations are acceptable in many applications. In these cases, the simplicity of the representation and the similarity to OBDDs makes the class of MTBDDs to an attractive data structure.

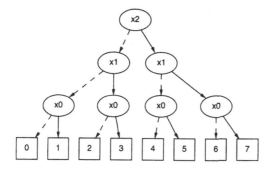

Figure 12.11. MTBDD representation of converting a binary number into the corresponding natural number

12.4.2 Edge Values

Typically, for applications where the number of admissible function values is large, MTBDDs are not suited. A way to obtain a suitable OBDD-based representation in theses cases arises from additionally introducing edge weights. By using these weights, sharing of common subgraphs can be improved.

Definition 12.25. *An edge-valued decision diagram (EVBDD) is defined like an OBDD, where each edge is additionally provided with an integer weight g.*

Each node v with label x_i computes a function $f_v : \mathbb{B}^n \to \mathbb{Z}$,

$$f_v = x(g_1 + f_1) + (1 - x)(g_0 + f_0),$$

where f_1, f_0 are the functions represented by the 1- and 0-successor of v, and g_1, g_0 are the corresponding edge weights, respectively. Additionally, to the function f being computed in the root node a fixed constant c can be added.

Example 12.26. The EVBDD in Fig. 12.12 describes the same function as the MTBDD in Example 12.11:

$$f(x_0, x_1, x_2) = x_0 + 2x_1 + 4x_2.$$

Here, the node with label x_0 represents the function f_0,

$$f_0 = x_0(0 + 1) + (1 - x_0)(0 + 0) = x_0.$$

The node with label x_1 represents the function f_1,

$$f_1 = x_1(2 + x_0) + (1 - x_1)(0 + x_0) = 2x_1 + x_0 x_1 + x_0 - x_1 x_0 = 2x_1 + x_0.$$

The MTBDD representation size of the function for converting an n-bit string into the corresponding natural number grows exponentially in n, but the growth of the EVBDD representation is linearly bounded. $\diamond$

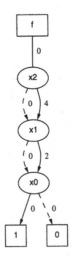

Figure 12.12. EVBDD representation for the conversion of a binary number into the corresponding natural number

By restricting the choice of admissible edge weights, EVBDDs can be brought into a canonical form, too. One possibility is to fix the edge value of each 0-edge to 0. In Fig. 12.12 this condition is already satisfied.

Interesting applications of the data structure of EVBDDs presented in the literature include solving problems which are tightly connected to integer optimization problems.

12.4.3 Moment Decompositions

Now we describe a representation form which is particularly suited for arithmetic functions of the type $f : \{0,1\}^n \to \mathbb{Z}$.

If one interprets switching functions as numerical functions with range $\{0,1\}$, then Shannon's expansion

$$f = x_i \cdot f_{x_i} + \overline{x_i} \cdot f_{\overline{x_i}}$$

can also be written in the form

$$f = x_i \cdot f_{x_i} + (1 - x_i) \cdot f_{\overline{x_i}}.$$

Here, $+$, $-$, and $\cdot$ denote the usual arithmetic operations over the integers. Of course, this decomposition is based on the precondition that the variable x_i can only assume the values 0 or 1.

The **moment decomposition** of a function is obtained by rearranging the terms in the above decomposition:

$$f = x_i \cdot f_{x_i} + (1 - x_i) \cdot f_{\overline{x_i}}$$
$$= f_{\overline{x_i}} + x_i \cdot (f_{x_i} - f_{\overline{x_i}}).$$

Definition 12.27. *Let $f \in \mathbb{B}_n$. For each variable x_i of f the function*

$$f_{\delta x_i} = f_{x_i} - f_{\overline{x_i}}$$

*is called the **linear moment** of f with respect to x_i.*

The terminology originates from the point of view which considers f as a linear function in its variables. Then $f_{\delta x_i}$ is the partial derivative of f with respect to x_i. As the function value of f is only defined for two values of x_i, we can always extend f to a linear function. Thus the moment decomposition is

$$f = f_{\overline{x_i}} + x_i \cdot f_{\delta x_i}.$$

Definition 12.28. *A **binary moment diagram** (BMD) is defined like an OBDD with one difference: the function $f : \mathbb{B}^n \to \mathbb{Z}$ represented by a BMD results from computing in each node the corresponding function according to the moment decomposition. Each node v with label x_i computes a function $f_v : \mathbb{B}^n \to \mathbb{Z}$,*

$$f_v = f_0 + x_i f_1,$$

where f_1, f_0 are the functions defined by the 1- and 0-successor of v, respectively.

Example 12.29. We consider again the conversion of a binary number into the corresponding natural number from Examples 12.11 and 12.12. A BMD of this function is shown in Fig. 12.13 (a).

Here, the node with label x_0 represents the function f_0,

$$f_0 = 0 + x_0 \cdot 1 = x_0,$$

and the node with label x_1 represents the function f_1,

$$f_1 = x_0 + x_1 \cdot 2 = x_0 + 2x_1.$$

The linear moment of the example function with respect to a variable x_i is exactly 2^i. In other words, the function is decomposed according to the individual bit significances. Due to this property BMDs are particularly suited to represent arithmetic functions.   ◇

Like MTBDDs, BMDs also provide a canonical representation. However, some of the relevant basic operations may imply exponential running times in the worst case. Although BMDs are typically quite efficient in practical

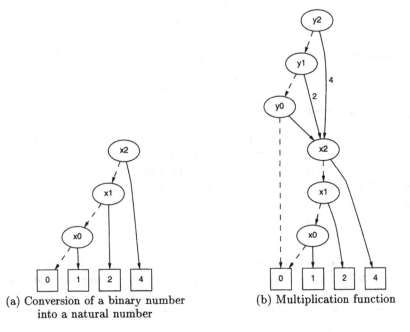

(a) Conversion of a binary number
into a natural number

(b) Multiplication function

Figure 12.13. BMDs and multiplicative BMDs

applications, it should always be checked whether their use can actually lead to significant size reductions with regard to the relevant functions.

A particularly interesting extension of BMDs is the concept of **multiplicative BMDs (*BMDs)**. These graphs also allow edge weights which serve as factor for the function that is represented at the target node of the edge. With help of suitable restrictions to the choice of edge weights, *BMDs can also be brought into a canonical form. By employing *BMDs it becomes possible to represent an n-bit multiplier $\{0,1\}^n \times \{0,1\}^n \mapsto \mathbb{Z}$ within *linear* space. Remember that due to Theorem 8.8 multipliers necessarily lead to exponential OBDDs on the bit level, and hence they cause serious problems in practice.

Example 12.30. Let $f : \{0,1\}^n \times \{0,1\}^n \mapsto \mathbb{Z}$ be the function whose input consists of two vectors $x = (x_{n-1}, \ldots, x_0)$ and $y = (y_{n-1}, \ldots, y_0)$, and which computes the natural number of their product

$$f(x,y) = \left(\sum_{i=0}^{n-1} x_i \, 2^i \right) \left(\sum_{j=0}^{n-1} y_j \, 2^j \right).$$

A *BMD in case $n = 3$ can be seen in Fig. 12.13 (b). Let f_{x_2}, f_{y_0} and f_{y_1} be the function which are represented in the nodes with label x_2, y_0 and y_1, respectively. Example 12.29 implies

$$f_{x_2}(x, y) = \sum_{i=0}^{2} x_i 2^i.$$

Hence,

$$f_{y_0}(x, y) = 0 + y_0 f_{x_2}(x, y) = y_0 \sum_{i=0}^{2} x_i 2^i,$$

$$f_{y_1}(x, y) = f_{y_0}(x, y) + 2y_1 f_{x_2}(x, y) = y_0 \sum_{i=0}^{2} x_i 2^i + 2y_1 \sum_{i=0}^{2} x_i 2^i$$

$$= \left(\sum_{i=0}^{2} x_i 2^i \right) \left(\sum_{j=0}^{1} y_j 2^j \right),$$

$$f(x, y) = f_{y_1}(x, y) + 4y_2 f_{x_2}(x, y) = \left(\sum_{i=0}^{2} x_i 2^i \right) \left(\sum_{j=0}^{1} y_j 2^j \right) + 4y_2 \sum_{i=0}^{2} x_i 2^i$$

$$= \left(\sum_{i=0}^{2} x_i 2^i \right) \left(\sum_{j=0}^{2} y_j 2^j \right).$$

For larger values of n this construction can be continued analogously.

If the represented circuit computes the multiplication function *correctly*, then, consequently, the representation in terms of a *BMD remains provably small. However, in industrial use of the verification tools, one can observe a very undesirable effect whose prevention is a topic of current research. If due to a design error a circuit C *does not* compute the multiplication function correctly, but instead, it differs merely very slightly, then it may happen that the size of the *BMD representation of C explodes. In other words, *BMDs are excellently suited to represent multipliers, but they are highly sensitive to small disturbances of the multiplier structure of the function.

12.5 References

The model of FBDDs was introduced as a data structure for Boolean manipulation by Gergov and Meinel [GM94a], and independently by Sieling and Wegener [SW95b]. The heuristic ideas for constructing complete types were presented in [BGMS94].

OFDDs were proposed by Kebschull, Schubert, and Rosenstiel [KSR92]; the combination of different decomposition types is due to Drechsler, Sarabi, Theobald, et al. [DST+94]. Applications of OFDDs in minimizing Reed-Muller expressions are investigated in [DTB96]. Recently, an alternative approach for constructing decision diagrams based on the EX-OR operation has been proposed in [MS98].

The variant of zero-suppressed BDDs goes back to Minato [Min93, Min96]. The theorem relating the sizes of OBDDs and ZDDs to each other was proven by Schröer and Wegener [SW95a]. Successful applications of ZDDs in the area of logic synthesis are presented in the survey [Cou94].

In the two papers [CMZ+93, BFG+93], the model of multi-terminal BDDs was introduced. The model of edge-valued BDDs was proposed in the paper [LPV94]. Finally, the moment decomposition, BMDs, and *BMDs were introduced and studied by Bryant and Chen [BC95].

13. Transformation Techniques for Optimization

Die Ringenden sind die Lebendigen.
[The struggling ones are the living.]
Gerhart Hauptmann (1862–1946): Der arme Heinrich

In this chapter, we introduce transformation techniques which serve to further optimize OBDD representations. The optimization space in this framework goes far beyond the optimization space established by the optimization of the variable order.

Domain transformations constitute a classic concept in mathematics, physics, and engineering. By transforming a function into a new space, where the representation of the function can be handled easily, applications can be improved and simplified. Examples of classic transformation concepts include the Fourier transformation, the Laplace transformation, and the Z-transformation. In this chapter, we explain how transformation techniques of this kind can also be successfully applied to the context of OBDD-based Boolean manipulation.

In contrast to the OBDD variants presented in Chapter 12, the use of domain transformations does not require a separate software package. Instead, it remains possible to work with conventional OBDD packages, merely the semantic interpretation of the OBDDs under consideration is changed.

13.1 Transformed OBDDs

In the transformation techniques discussed below, the concept of *cube transformations* stands in the center of attraction. These cube transformations describe a repositioning of the vertices of a Boolean cube, which induces a transformation on the argument domain of the switching function that has to be represented.

Definition 13.1. *A* **cube transformation** τ *is a bijective mapping* $\tau : \mathbb{B}^n \rightarrow \mathbb{B}^n$.

A cube transformation τ defines a mapping $\phi_\tau : \mathbb{B}_n \to \mathbb{B}_n$ onto the Boolean algebra of all switching functions in n variables through

$$\phi_\tau(f)(a) = f(\tau(a))$$

for all $a = (a_1, \ldots, a_n) \in \mathbb{B}^n$. ϕ_τ is called the mapping **induced** by τ. The following lemma can be immediately deduced from the definition of a cube transformation.

Lemma 13.2. *For each cube transformation $\tau : \mathbb{B}^n \to \mathbb{B}^n$, the mapping ϕ_τ defines an automorphism on $\mathbb{B}_n$, i.e., for all functions $f, g \in \mathbb{B}_n$ we have:*

1. $f = g$ if and only if $\phi_\tau(f) = \phi_\tau(g)$.

2. For any binary operation $*$ on $\mathbb{B}_n$ it holds $\phi_\tau(f * g) = \phi_\tau(f) * \phi_\tau(g)$. $\square$

In other words: the second property says that it is not important whether first a Boolean operation is applied to the functions f and g and then the resulting function is transformed, or whether first the functions f and g are transformed and then the transformation is applied. Both ways lead to the same result. If the operation is performed in terms of OBDDs, then the polynomial complexity is ensured also when working with the transformed functions. Now let us consider the equivalence test of two function representations in a situation where only the transformed representations are available. Then the first statement of the lemma tells us that in this situation it is not necessary at all to retransform the functions. As ϕ_τ is bijective, the decision is also possible on the transformed representations.

According to the following definition the OBDDs of transformed switching functions are denoted as *TBDDs*.

Definition 13.3. *Let $f \in \mathbb{B}_n$, and let τ be a cube transformation $\tau : \mathbb{B}^n \to \mathbb{B}^n$. A τ-**TBDD** representation of f is an OBDD representation of $\phi_\tau(f)$.*

If we succeed in finding and constructing those transformations τ which lead to small τ-TBDD representations of the relevant switching functions, then OBDD-based algorithms can be substantially accelerated by applying them on the transformed switching functions.

Example 13.4. Let $f(x_1, x_2, x_3) = \overline{x_1}x_2 + x_1x_3$. Figure 13.1 shows a transformation $(y_1, y_2, y_3) = \tau(x_1, x_2, x_3)$ in explicit form. On the right side of Fig. 13.1, the OBDDs of the two functions f and $\phi_\tau(f)$ are depicted. The function f computes a 1 for an input (x_1, x_2, x_3) if and only if

$$(x_1, x_2, x_3) \in \{(0, 1, 0), (0, 1, 1), (1, 0, 1), (1, 1, 1)\}.$$

(x_1, x_2, x_3)	$\tau(x_1, x_2, x_3) = (y_1, y_2, y_3)$
$(0,0,0)$	$(0,0,0)$
$(0,0,1)$	$(0,0,1)$
$(0,1,0)$	$(0,1,0)$
$(0,1,1)$	$(0,1,1)$
$(1,0,0)$	$(1,0,0)$
$(1,0,1)$	$(1,1,0)$
$(1,1,0)$	$(1,0,1)$
$(1,1,1)$	$(1,1,1)$

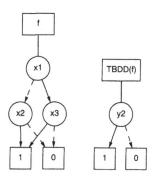

Figure 13.1. Example of a TBDD representation

According to the transformation τ, the function $\phi_\tau(f)$ computes a 1 for an input $(y_1, y_2, y_3) = \tau(x_1, x_2, x_3)$ if and only if

$$(y_1, y_2, y_3) \in \{(0,1,0), (0,1,1), (1,1,0), (1,1,1)\}.$$

Hence, $\tau(f) = y_2$. When applying a cube transformation the size of the on-set remains invariant, but a reduction of memory consumption can be achieved nevertheless. In our example, the OBDD of f requires three internal nodes, but a single node suffices for the OBDD of $\phi_\tau(f)$.   ◇

The well-known properties of OBDDs in connection with the fact that cube transformations induce automorphisms on $\mathbb{B}_n$ (see Lemma 13.2) immediately imply the following properties of TBDDs.

Theorem 13.5. *Let $f, f_1, f_2 \in \mathbb{B}_n$ be switching functions, and let $\tau : \mathbb{B}^n \to \mathbb{B}^n$ be a cube transformation. Further, let π be an order on the set of variables $\{x_1, \ldots, x_n\}$.*

Universality of TBDD representation: *Any switching function $f \in \mathbb{B}_n$ can be represented by a (reduced) τ-TBDD with respect to the order π.*

Canonicity of TBDD representation: *Any switching function $f \in \mathbb{B}_n$ has exactly one representation as reduced τ-TBDD with respect to the order π.*

Efficient synthesis of TBDDs: *Let T_1, T_2 be two τ-TBDD representations of f_1 and f_2 with respect to π. For each binary Boolean operation $*$ the (reduced) τ-TBDD representation of $f = f_1 * f_2$ with respect to π can be computed in time $\mathcal{O}(size(T_1) \cdot size(T_2))$.*

Efficient equivalence test: *Let T_1 and T_2 be the reduced τ-TBDDs of f_1, f_2 with respect to π. Equivalence of the functions f_1 and f_2 corresponds to the functional equivalence of their representations T_1 and T_2, and can be tested in linear time. In case of a strongly canonical representation only constant time is necessary.*   □

The universality of the TBDD representation says that each switching function can also be represented by a TBDD. However, this fact alone does not suffice to make TBDDs an interesting representation type, as already the model of OBDD provides a universal and unique representation of switching functions. The real aim of the concept of TBDDs is to reduce the memory consumption during the manipulation process significantly, while keeping the well-understood "manipulation language" of OBDDs. Indeed, it can be shown that for each switching function f there exists a cube transformation τ such that f has a very small τ-TBDD representation.

Theorem 13.6. *For each switching function $f \in \mathbb{B}_n$ there exists a τ-TBDD representation of size n.*

Proof. Let $k = |\text{on}(f)|$. We consider the inputs $a = (a_1, \ldots, a_n) \in \mathbb{B}^n$ as binary representations of the numbers $0, \ldots, 2^n - 1$. Let τ be the bijection which maps those strings representing the numbers $0, \ldots, k - 1$ onto the elements of $\text{on}(f)$. Then $\phi_\tau(f)$ can be represented by an OBDD with natural variable order which tests whether the input (interpreted as binary number) is smaller then k (in this case the 1-sink is reached) or not (in this case the 0-sink is reached). This test can be performed by an OBDD which for each variable x_i has at most one node labeled by x_i.   □

Before this result is celebrated euphorically as solution to all representation problems, one should be aware of the fact that still some problems have to be coped with. In particular, it may be very difficult and costly to store and to manipulate the mentioned transformation for specific functions.

To conclude this introductory section on TBDDs we would like to describe how combinational circuit verification can be realized in terms of TBDDs. Starting from two net lists of gates it is to check for the circuits C and C' whether C and C' compute the same functions. When applying OBDD-based methods we start from the (trivial) OBDD representations of the primary input variables $x_1, \ldots, x_n$ and successively construct OBDDs for each of these gates from the OBDDs of the predecessor gates. After this symbolic simulation of the two circuits, a comparison of the resulting OBDDs – realized by a single pointer comparison – suffices to determine whether the two net lists are functionally equivalent. For very complex circuits it may happen that during constructing the OBDDs the available memory is exceeded, and therefore the OBDD representation of the two circuits cannot be constructed at all.

Now the TBDD concept allows us to work with a suitably transformed function instead of the original function. In order to check whether two circuits C and C' are equivalent, it suffices to show that two reduced TBDD representations T_C and T'_C of C and C' coincide. As the mapping ϕ_τ on $\mathbb{B}_n$, which is induced by the cube transformation τ, is an automorphism, the desired

TBDD representation can be constructed in exactly the same way as in case of OBDDs. First, one generates the τ-TBDD representation of the variables $x_1, \ldots, x_n$, and then computes (now within a usual OBDD environment) the τ-TBDDs of C and C' by means of a symbolic simulation. With the exception of the construction of the TBDDs of the variables $x_1, \ldots, x_n$, all steps can be performed by employing a conventional OBDD package.

When optimizing OBDDs by means of cube transformations, the main task is to characterize "feasible" classes of cube transformations from the tremendously large set of $(2^n)!$ different cube transformations in n variables. Here, "feasible" means that the cube transformations in the chosen class can be handled efficiently enough to be suitable for practical applications.

Practical classes of cube transformations have to satisfy primarily two conditions:

- It must be possible to store the transformations τ efficiently.
- It must be possible to compute the τ-TBDD representations of the primary input variables easily.

In the following sections, we treat two classes of cube transformations which satisfy these two properties, namely type-based transformations and linear transformations.

13.2 Type-Based Transformations

13.2.1 Definition

One way to obtain easy-to-handle but nevertheless powerful cube transformations is to apply complete types, as introduced in Definition 12.1.

By using complete types, cube transformations can be defined as follows. Throughout the description, let σ be a fixed chosen complete type. Then each vector $a = (a_1, \ldots, a_n) \in \mathbb{B}^n$ defines a uniquely determined path $p(a)$ from the root to the sink of σ. $a^{[i]}$ denotes the index of the variables which are tested in the i-th position of $p(a)$. Now the cube transformation $\tau_\sigma : \mathbb{B}^n \to \mathbb{B}^n$ which is induced by the complete type σ is defined by

$$\tau_\sigma(a_1, \ldots, a_n) = (a_{a^{[1]}}, \ldots, a_{a^{[n]}}).$$

Before proving that τ_σ is actually bijective and hence a cube transformation, we first consider an example.

Example 13.7. For the complete type in Fig. 13.2 we consider the input $a = (a_1, a_2, a_3, a_4) = (1, 0, 0, 0)$. The path which is induced by the input is

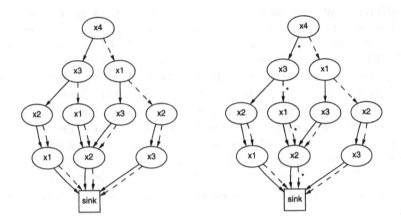

Figure 13.2. The left side shows a complete type of the variables x_1, x_2, x_3, x_4. In the right picture, additionally the path of the input $(1, 0, 0, 0)$ is marked

marked in the right part of Fig. 13.2. On this path, the variables are traversed in the sequence x_4, x_3, x_1, x_2. Hence,

$$\tau_\sigma(1,0,0,0) = (a_4, a_3, a_1, a_2) = (0,0,1,0).$$

$\Diamond$

Theorem 13.8. *Let σ be a complete type over the set of variables $\{x_1, \ldots, x_n\}$. Then τ_σ defines a cube transformation on $\mathbb{B}^n$.*

Proof. As σ is a complete type and, hence, each variable occurs on each path from the root to the sink, the mapping τ_σ is completely defined. In order to show that τ_σ is a cube transformation, it therefore suffices to show that τ_σ is injective. Let $a, b \in \mathbb{B}^n$ with $\tau_\sigma(a) = \tau_\sigma(b)$. Then, by definition of τ_σ, the complete type σ satisfies

$$a_{a[i]} = b_{b[i]}, \quad 1 \le i \le n.$$

First, we show that $a^{[i]} = b^{[i]}$ for all $1 \le i \le n$. Since, by Definition 12.1, we have $c^{[1]} = d^{[1]}$ for any $c, d \in \mathbb{B}^n$, the equation $a_{a[1]} = b_{b[1]}$ implies $a^{[2]} = b^{[2]}$. Analogously, $a_{a[2]} = b_{b[2]}$ implies $a^{[3]} = b^{[3]}$. Inductively, we obtain $a_{a[i]} = b_{b[i]}$ for all $1 \le i \le n$.

Now let $j \in \{1, \ldots, n\}$. Due to the read-once property of a complete type there is exactly one $i \in \{1, \ldots, n\}$ with $a^{[i]} = j$. Hence,

$$a_j = a_{a[i]} = b_{b[i]} = b_{a[i]} = b_j.$$

As this statement is true for all j, we haven proven $a = b$. Consequently, τ_σ is injective. $\square$

Note that linear orders only permute the set of variables, i.e., the indices in the arguments. Hence, the space of cube transformations which are induced by linear orders is isomorphic to the optimization space which is generated by the choice of a variable order. However, by means of complete types much more complex cube transformations can be defined. Here, the indices of the arguments can be permuted depending on the arguments themselves.

13.2.2 Circuit Verification

The framework of using type-based TBDDs as tool for combinational circuit verification was already described in Section 13.1. It was based on symbolic simulation, where in the first step the transformed functions on the input variables have to be computed. Now we discuss how to realize this first step for the class of type-based transformations.

Let C be a circuit, and let σ be a complete type over the set of variables $\{x_1, \ldots, x_n\}$ which underlies the TBDD-representation of C. The idea behind the construction of the reduced τ_σ-TBDDs P_i of the variables x_i, i.e., the reduced OBDDs of the switching functions resulting from the variables x_i through the transformation τ_σ, is as follows. We consider a fixed $i \in \{1, \ldots, n\}$. For each input $(b_1, \ldots, b_n) \in \mathbb{B}^n$, the OBDD P_i has to compute a 1 if and only if $a_i = 1$ for $a = \tau_\sigma^{-1}(b)$.

Starting from the type σ, all nodes below the nodes with label x_i are removed. We connect the 1-edge of each node labeled by x_i with a newly introduced 1-sink, and the 0-edge of each node labeled by x_i with a newly introduced 0-sink. Then we change the labels of the variables. The root obtains the label y_1, all successors of the root the label y_2, and so on.

Now let $b = (b_1, \ldots, b_n) \in \mathbb{B}^n$ be an input, and let $a = (a_1, \ldots, a_n) = \tau_\sigma^{-1}(b) \in \mathbb{B}^n$. Further, let j be the position of the variable y_i on the path in P_i which is induced by b. The input b yields an output value 1 if and only if $b^{[j]} = 1$. Since $b = \tau_\sigma(a)$, this is the case if and only if $a_{a^{[j]}} = 1$. By definition of j we have $a^{[j]} = i$. Hence, b implies an output value 1 in P_i, if for any $a = \tau_\sigma^{-1}(b)$ it holds $a_i = 1$.

Finally, the constructed OBDD P_i can be reduced. Figure 13.3 shows pseudo code to realize this idea. If already computed subproblems are stored, then in the course of performing this algorithm each node has to be considered only once. In the pseudo code of the algorithm this is indicated by the function *set_mark*.

Example 13.9. We illustrate the algorithm in Fig. 13.3 with the example of the variable x_3 of the complete type σ in Fig. 13.2. All nodes below the nodes with label x_3 are removed, and the two outgoing edges of the nodes with label x_3 are connected with the newly introduced sinks. Figure 13.4 shows this intermediate step and the τ_σ-TBDD of the variable x_3 in the transformed variables $y_1, \ldots, y_4$, where this TBDD is in reduced form.   ◇

```
transform_variables(σ, i) {
/* Input: A complete type σ and an integer i ∈ {1, . . . , n} */
/* Output: A τ_σ-OBDD of x_i over the (transformed) set of variables
            {y_1, . . . , y_n} */
    transform_step(i, 1, {x_1, . . . , x_n}, σ);
    clear_mark_below(σ);
}



transform_step(σ, i, M, t) {
    If M = ∅ Return Undefined;
    If marked(t) {
        Return result(t);
    }
    set_mark(t);
    Let the node t be described by the triple (x, t_1, t_0);
    If x_i ∈ M \ {x}
        f_1 = transform_step(i, r + 1, M{x}, t_1);
        f_0 = transform_step(i, r + 1, M{x}, t_0);
        reduce_and_return(y_r, f_1, f_0);
    Else {
        If x_i ≠ x Then Return Undefined;
        Else reduce_and_return(y_r, 1, 0);
    }
}
```

Figure 13.3. Algorithm for constructing the OBDDs of the transformed variables

A complexity analysis of the algorithm yields the following estimation:

Theorem 13.10. *Let σ be a complete type over the set of variables $\{x_1, \ldots, x_n\}$, and let P_i be the reduced τ_σ-TBDDs of the variables x_i, $1 \leq i \leq n$. Each P_i has at most as many nodes as σ and can therefore be computed in linear time and linear space with respect to the size of σ.* □

Of course, a practical realization of this concept not only needs these basic algorithms but also requires heuristics that automatically deduce good types from the structure of a given problem. For the case of symbolic simulation, J. Bern, Ch. Meinel, and A. Slobodová have proposed heuristics for constructing good cube transformations from a circuit topology. Comparing these TBDD heuristics with the well-known heuristics for constructing good variable orders of OBDDs, the TBDD-based approach leads to significantly smaller representations. In some specific cases, the TBDD-based approach succeeds in symbolic simulating circuits whose representation by a "classical" OBDD exceeds the available memory.

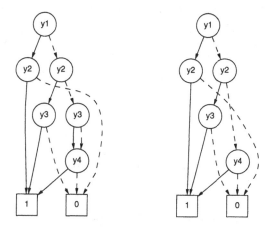

Figure 13.4. Intermediate step and reduced τ_σ-TBDD of the variable x_3

13.3 Linear Transformations

13.3.1 Definition

In this section, we go into a transformation concept which can also be applied dynamically and hence forms the counterpart to dynamic reordering algorithms. The concept is based on linear transformations.

Definition 13.11. *An* **elementary linear transformation** $\tau_{i,j} : \mathbb{B}^n \to \mathbb{B}^n$, $1 \le i, j \le n$, $i \ne j$, *is defined by the following mapping:*

$$\tau_{i,j}(x_1, \dots, x_n) = (x_1, \dots, x_{i-1}, x_i \oplus x_j, x_{i+1}, \dots, x_n).$$

For this mapping, we write $x_i \mapsto x_i \oplus x_j$.

Obviously, the mapping $\tau_{i,j}$ is a one-to-one correspondence and therefore a cube transformation.

Considering the set $\mathbb{B}^n$ together with the two Boolean operations $\oplus$ and $\cdot$ as a field, this transformation is a linear mapping in the algebraic sense, i.e., there is an $n \times n$-matrix A with entries in $\mathbb{B}$ such that

$$\tau_{i,j}(x_1, \dots, x_n)^T = A \cdot (x_1, \dots, x_n)^T.$$

Here, $(x_1, \dots, x_n)^T$ denotes the transposed vector to $(x_1, \dots, x_n)$, and the matrix product is computed with respect to the operations $\oplus$ and $\cdot$. In case of the elementary linear transformation $\tau_{i,j}$, besides the diagonal elements, only the rows i and j as well as the columns i and j may contain non-zero entries.

Example 13.12. We consider a Boolean function in 3 variables x_1, x_2 and x_3. By the linear transformation $x_2 \mapsto x_2 \oplus x_3$ the input vector $(0, 1, 1)$ is mapped to the vector $(0, 0, 1)$. The matrix A corresponding to this transformation is

$$A = \begin{pmatrix} 1\ 0\ 0 \\ 0\ 1\ 1 \\ 0\ 0\ 1 \end{pmatrix}.$$

$\diamond$

Of course, each elementary linear transformation $x_i \mapsto x_i \oplus x_j$ can be described by the following truth table in the variables x_i and x_j. For large n, this representation is much more compact than the matrix representation.

(x_i, x_j)		$(x_i \oplus x_j, x_j)$	
0	0	0	0
0	1	1	1
1	0	1	0
1	1	0	1

From linear algebra, it is well-known that the set of elementary linear transformations constitutes a generating system of the set of all bijective (i.e. invertible) linear transformations. This implies that any given invertible linear transformation can be generated by a sequence of elementary linear transformations. The following lemma already indicates that the optimization space induced by linear transformations is quite suitable for minimizing OBDD representations. The space is substantially larger than the optimization space of different variable orders. However, it is still much smaller than the number of all cube transformations, and thus it seems possible to gain control of the class of linear transformations.

Lemma 13.13. *(1) The number $t(n)$ of invertible linear transformations on $\mathbb{B}^n$ amounts to*

$$t(n) = \prod_{i=0}^{n-1} (2^n - 2^i).$$

(2) The quotient of $t(n)$ and the number $(2^n)!$ of all cube transformations on $\mathbb{B}^n$ converges to 0, as n tends to infinity.

(3) The quotient of the number $n!$ of all possible variable orders in n variables and the the number $t(n)$ converges to 0, as n tends to infinity.

Proof. (1) The number of invertible linear transformations on $\mathbb{B}^n$ coincides with the number of regular $n \times n$-matrices over the field $\mathbb{Z}_2$. The number of these matrices can be computed as follows. The first row vector b_1 can be chosen arbitrarily from the set $\mathbb{Z}_2^n \setminus \{0\}$. The i-th row vector b_i, $2 \leq i \leq n$, can be chosen arbitrarily from the set of vectors which are linearly independent from $\{b_1, \ldots, b_{i-1}\}$ over $\mathbb{Z}_2^n$,

$$b_i \in \mathbb{Z}_2^n \setminus \left\{ \sum_{j=1}^{i-1} \lambda_j b_j \ : \ \lambda_1, \ldots, \lambda_{i-1} \in \mathbb{Z}_2 \right\}.$$

Hence, for choosing the vector b_i there are $2^n - 2^{i-1}$ possibilities. This proves the claim.

(2) It holds

$$\frac{t(n)}{(2^n)!} \leq \frac{2^{n^2}}{(2^n)!} \leq \frac{(n^2)!}{(2^n)!} \to 0 \ \text{ as } \ n \to \infty.$$

(3) It holds

$$\frac{(2n)!}{t(n)} = \frac{2n(2n-1)}{2^n - 2^0} \cdot \frac{(2n-2)(2n-3)}{2^n - 2^1} \cdots \frac{2 \cdot 1}{2^n - 2^{n-1}} \to 0 \ \text{ as } \ n \to \infty.$$

$\square$

13.3.2 Efficient Implementation

Before explaining how linear transformations can be employed for optimizing OBDD-based data structures in an efficient way, let us recall the variable swap of neighboring variables in the order, presented in Chapter 9. This time, our viewpoint is slightly different and serves for the incorporation of linear transformations. The process of swapping two neighboring variables in the order can be interpreted as cube transformation. We assume that the variable x_i appears in the order immediately before the variable x_j. The effect of a variable swap on a node with label x_i can be seen by applying Shannon's expansion,

$$f = x_i x_j f_{11} + x_i \overline{x_j} f_{10} + \overline{x_i} x_j f_{01} + \overline{x_i}\, \overline{x_j} f_{00}. \tag{13.1}$$

Now we consider the cube transformation τ which maps two variables x_i and x_j to each other,

$$\tau(x_1, \ldots, x_n) = (x_1, \ldots, x_{i-1}, x_j, x_{i+1}, \ldots, x_{j-1}, x_i, x_{j+1}, \ldots, x_n).$$

The function $\phi_1(f) = \phi_\tau(f)$ induced by this cube transformation results from the same effect as the one that is caused by swapping the variables x_i and x_j in the order. Rearranging the individual terms in $\phi_1(f)$, such that x_i occurs before x_j in each product, yields

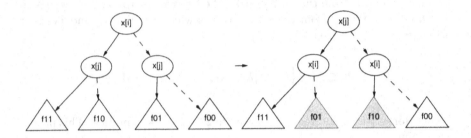

Figure 13.5. Swapping two neighboring variables x_i, x_j

$$\phi_1(f) = x_i x_j f_{11} + x_i \overline{x_j} f_{01} + \overline{x_i} x_j f_{10} + \overline{x_i}\,\overline{x_j} f_{00}. \tag{13.2}$$

In other words, the effect of a variable swap is to interchange f_{10} and f_{01}. Figure 13.5 illustrates this effect.

If we repeat the same process for the application of a linear transformation $x_i \mapsto x_i \oplus x_j$, then the resulting function $\phi_2(f)$ is

$$\phi_2(f) = x_i x_j f_{01} + x_i \overline{x_j} f_{10} + \overline{x_i} x_j f_{11} + \overline{x_i}\,\overline{x_j} f_{00}. \tag{13.3}$$

A comparison of the two Equations (13.2) and (13.3) shows that there is only a single difference: in case of the linear transformation, instead of the two functions f_{10} and f_{01}, the two functions f_{11} and f_{01} are interchanged. Hence, the fundamental techniques for efficiently implementing a swap of two variables can also be applied for their linear combination. Figure 13.6 illustrates the described elementary transformation.

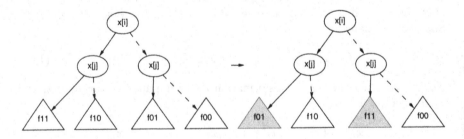

Figure 13.6. $x_i \mapsto x_i \oplus x_j$ for two neighboring variables x_i, x_j

Another inspection of the Equations (13.1) to (13.3) shows: by means of the variable swap, the linear transformation, and the original state, all partitions

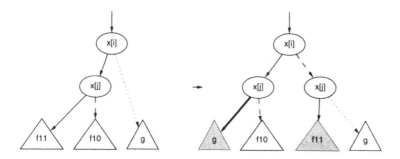

Figure 13.7. In case of complemented edges, locality of the transformation $x_i \mapsto x_i \oplus x_j$ is lost

of the set $\{f_{11}, f_{10}, f_{01}, f_{00}\}$ in two subsets of equal size can be generated. Hence, by combining variable swapping and linear transformations it is possible to reduce the size of the OBDDs in more cases than by variable swapping only.

In case of complemented edges, another difficulty has to be take into account. For this, we consider the OBDD with complemented edges from Fig. 13.7. The 0-edge of the node with label x_i is complemented and points to a sub-OBDD g whose root is not labeled by x_j. The linear transformation $x_i \mapsto x_i \oplus x_j$ now causes the 1-edge of the left x_j-node to become complemented, which is indicated by a bold arrow. In order to re-establish canonicity, the complement bit has to be moved up at least to a position above the node labeled by x_i. This fact destroys the locality of the linear transformation, a highly undesirable property.

If instead of the linear transformation $x_i \mapsto x_i \oplus x_j$ the complementary operation $x_i \mapsto x_i \equiv x_j$, i.e., the equivalence operation, is applied, then the problem vanishes. Figure 13.8 illustrates the transformation $x_i \mapsto x_i \equiv x_j$. Now the two functions f_{00} and f_{10} are interchanged. Indeed, nothing is lost by this modification; the application of the equivalence operation also allows one to generate all partitions of the set $\{f_{11}, f_{10}, f_{01}, f_{00}\}$ in two subsets of equal size.

13.3.3 Linear Sifting

As the class of linear transformations on neighboring variables can be implemented efficiently, this class provides a tool which offers further optimization potential with regard to the minimization of OBDDs. However, once more the question arises of how to convert this basic concept into an automatic

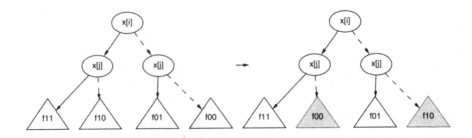

Figure 13.8. Transformation $x_i \mapsto x_i \equiv x_j$ for two neighboring variables x_i, x_j

procedure. A promising approach in this context is to combine the two basic operations on neighboring variables, namely

- the variable swap and
- the linear combination of two variables

within a single optimization algorithm.

To realize these ideas, Ch. Meinel, F. Somenzi, and T. Theobald proposed integrating linear transformations into what is currently the best reordering method: the sifting algorithm. The resulting algorithm has obtained the name **linear sifting**.

As in the conventional sifting procedure, in linear sifting each variable is moved once through the whole order. Let x_i be the currently considered variable, and let x_j be the immediate successor of x_i. The basic step of the sifting algorithm consists in swapping the two variables in the order. In contrast to this, the basic step of the linear sifting algorithm consists of the following three phases:

1. The variables x_i and x_j are swapped in the order. Let s_1 be the size of the OBDD after this swapping process.

2. The linear transformation $x_i \mapsto x_i \equiv x_j$ is applied. Let s_2 be the size of the resulting OBDD.

3. If $s_1 \leq s_2$, then the linear transformation is immediately undone. This is achieved by simply applying the transformation again, since it is its own inverse.

In each basic step the variable x_i is moved one position onward in the order and possibly linearly combined with the variable x_j.

Of course, an algorithm that reorders variables in an OBDD has to keep track of the permutation produced. Likewise, the linear sifting algorithm has to keep track of the generated linear transformation. The most expedient way to do that is to maintain the transformation itself within the shared OBDD. When applying a transformation to a shared OBDD, the OBDDs of the original literals x_i are also transformed. In this way, the generated linear transformation is automatically carried along.

In order to make the performance of the linear sifting competitive with the sifting algorithm, the ideas for improving efficiency from Section 9.2.4 have also to be taken into account. A central device in an efficient sifting realization is the interaction matrix which says whether there is a function in the OBDD which simultaneously depends on two given variables x_i and x_j. If two variables are not interacting in this sense, then the swap can be performed in constant time. The interaction matrix is also effective in the case of linear sifting. On the one hand, it allows fast swaps in case of two non-interacting variables. On the other hand, it can be used to avoid linear combinations of variables that do not interact. In contrast to the exclusive application of the swap operation, when linearly combining two variables the interaction matrix can change.

Example 13.14. We consider the case of two functions in three variables:

$$f = x \oplus y,$$
$$g = y \equiv z.$$

The variables x and y do not interact, as neither f nor g depends simultaneously on both variables. However, the transformation $y \mapsto x \equiv y$ generates the transformed functions

$$\phi(f) = \overline{y},$$
$$\phi(g) = x \oplus y \oplus z.$$

Now all variables interact with each other. $\diamond$

If during performing the linear sifting algorithm, two variables are combined with each other, then the interaction matrix is updated accordingly.

Due to the additional costs of the linear transformations within a basic step, the time consumption of linear sifting is by a factor of 2 to 3 larger than the time consumption of sifting. Experimental studies with regard to the size of the resulting graphs have shown that in many cases the use of linear sifting may lead to significantly smaller representations compared to the classical sifting algorithm. In extreme cases, such as the circuits C499 and C1355 (see Section 9.3), the OBDD size obtained after an optimization by means of variable reordering can be reduced by more than 90% when using linear sifting.

13.4 Encoding Transformations

A central element of finite state machines, as introduced in Section 2.7, is
the state set Q. Before the system can be analyzed by means of Boolean
manipulation techniques or synthesized into a digital circuit, each state has to
be identified with a binary bit-string. This process is called **state encoding**.
An example can be seen in Fig. 13.9.

Symbolic state	Encoding
q_0	00
q_1	01
q_2	10
q_3	11

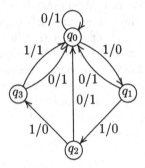

Figure 13.9. State encoding

The actual input/output behavior of the system is not influenced at all by
the choice of the concrete encoding. However, the encoding can strongly
influence the complexity of the manipulation of the system. In connection
with representation and analysis of sequential systems the state encoding can
therefore be applied as an additional optimization parameter. In order to
demonstrate how strongly the size of the OBDD representation may depend
on the choice of the state encoding, we consider finite state machines with a
very simple structure.

Definition 13.15. *An **autonomous counter** with 2^n states $q_0, \dots, q_{2^n-1}$
is an autonomous (i.e., input-independent) finite state machine with $\delta(q_i) =
q_{i+1}$, $0 \leq i < 2^n - 1$, $\delta(q_{2^n-1}) = q_0$.*

Such a counter occurs – at least as a submodule – in many sequential systems. Figure 13.10 illustrates the structure of the autonomous counter. The
following theorem shows that almost all encodings of the autonomous counter
lead to OBDD representations of exponential size, even with respect to the
optimal order.

Theorem 13.16. *Let $G(n)$ be the number of n-bit encodings of the autonomous counter with 2^n states which have a (shared) OBDD size of at most
$2^n/n$ with respect to their optimal order. Further, let $N(n) = (2^n)!$ be the
number of all possible n-bit counter encodings. Then the quotient $G(n)/N(n)$
converges to 0, as n tends to infinity.*

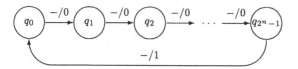

Figure 13.10. Autonomous counter

Proof. Let $k = \lfloor 2^n/n \rfloor$. Analogous to the proof of Theorem 8.3 one can show that there exist at most

$$n! \binom{n+k}{k} ((k+1)!)^2 = (k+1)(k+1)!(k+n)!$$

shared OBDDs with at most k nodes with respect to their optimal order. Due to the cyclic symmetry of the counter we have

$$\begin{aligned}
G(n) &\le 2^n\, n!\, (n+k)!\, ((k+1)!)^2 \\
&\le 2^n\, (2n+3k+2)! \\
&\le 2^n\, (2n+3\cdot 2^n/n+2)! \\
&\le 2^n\, (4\cdot 2^n/n)! \quad \text{for sufficiently large } n,
\end{aligned}$$

and hence,

$$\frac{G(n)}{N(n)} \to 0 \quad \text{as } n \to \infty.$$

$\square$

Definition 13.17. *An* **encoding transformation** *or a* **re-encoding** *is a bijective mapping* $\varrho : \mathbb{B}^n \to \mathbb{B}^n$, *which converts a given state encoding to a new encoding. If a state s is encoded by a bit-string $c \in \mathbb{B}^n$, then its new encoding is $\varrho(c)$.*

An example of a re-encoding can be seen in Fig. 13.11. The change in the internal state encoding does not modify the input/output behavior of the sequential system. Let the system with the new encoding be denoted by $M' = (Q', I, O, \delta', \lambda', q_0')$. The encoded next-state function, output function and the encoded initial state of M' can be computed as follows:

$$\begin{aligned}
\delta'(s,e) &= \varrho(\delta(\varrho^{-1}(s),e)), \\
\lambda'(s,e) &= \lambda(\varrho^{-1}(s),e), \\
q_0' &= \varrho(q_0),
\end{aligned} \tag{13.4}$$

where $\delta, \delta' : \mathbb{B}^n \times \mathbb{B}^p \to \mathbb{B}^n$, $\lambda, \lambda' : \mathbb{B}^n \times \mathbb{B}^p \to \mathbb{B}^m$, and $q_0, q_0' \in \mathbb{B}^n$.

	Old encoding	New encoding
q_0	00	01
q_1	01	11
q_2	10	00
q_3	11	10

Figure 13.11. Re-encoding of the symbolic states $q_0, \dots, q_3$

The transition relation of the re-encoded machine M' can be obtained from the transition relation of M as follows:

Lemma 13.18. *Let $T(x, y, e)$ be the characteristic function of the transition relation of M. Then the characteristic function $T'(x, y, e)$ of the transition relation of M' is*

$$\prod_{i=1}^{n} \left(\varrho_i^{-1}(y) \equiv \delta_i(\varrho^{-1}(x)) \right).$$

Therefore $T'(x, y, e)$ can be obtained from $T(x, y, e)$ by the substitutions $y_i \mapsto \varrho_i^{-1}(y)$ and $x_i \mapsto \varrho_i^{-1}(x)$, $1 \leq i \leq n$.

Proof. The claim is a consequence of the following equivalences:

$$
\begin{aligned}
T'(x, y, e) = 1 &\iff y_i = \varrho_i(\delta(\varrho^{-1}(x))) \text{ for all } i \\
&\iff y = \varrho(\delta(\varrho^{-1}(x))) \\
&\iff \varrho^{-1}(y) = \delta(\varrho^{-1}(x)) \\
&\iff \prod_{i=1}^{n} \left(\varrho_i^{-1}(y) \equiv \delta_i(\varrho^{-1}(x)) \right) = 1.
\end{aligned}
$$

$\square$

The optimization potential of encoding transformations can now be illustrated at the example of the autonomous counter. First we show that there exist encodings whose transition relation is very small, even if the variable order is fixed to $x_1, y_1, \dots, x_n, y_n$. We consider the **standard encoding** of the counter, where the encoding of the state q_i is the binary representation of i (see Fig. 13.12). Let M_{2^n} be the autonomous counter with n states under this encoding.

State	Encoding
q_0	$00\ldots000$
q_1	$00\ldots001$
q_2	$00\ldots010$
$\vdots$	$\vdots$
q_{2^n-1}	$11\ldots111$

Figure 13.12. Standard encoding

Lemma 13.19. *For $n \geq 2$, the reduced OBDD of the transition relation of M_{2^n} with respect to the variable order $x_1, y_1, \ldots, x_n, y_n$ consists of exactly $5n - 1$ nodes.*

Proof. The idea is based on using the OBDD of the transition relation of $M_{2^{n-1}}$, in order to construct the OBDD of the transition relation of M_{2^n}. Formally, this leads to a proof by induction. We show: the reduced OBDD has the form shown in Fig. 13.13 (a) (in the sense that the depicted nodes exist and are not pairwise isomorphic) with sub-OBDDs A and B and contains exactly $5n - 1$ nodes. The case $n = 2$ can be checked easily.

(a) Structure   (b) Induction   (c) Induction step
   of the OBDD   hypothesis

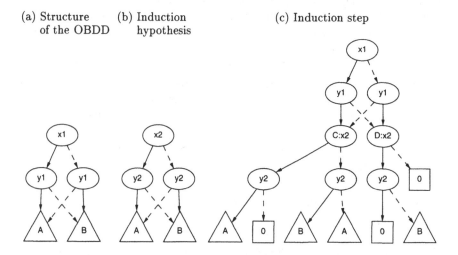

Figure 13.13. Inductively constructing the OBDD of the transition relation of the autonomous counter

Induction step: The OBDD for the $n - 1$ bits $x_2, \ldots, x_n$ has the form shown in Fig. 13.13 (b). Let $|x|$ be the binary number defined by a bit-string x. With this notation we have for the roots of the sub-OBDDs A, B:

A: leads to the 1-sink if and only if $|y_3 \ldots y_n| = |x_3 \ldots x_n| + 1$, $x_3 \ldots x_n \neq 11 \ldots 1$.

B: leads to the 1-sink if and only if $x_3 = \ldots = x_n = 1$, $y_3 = \ldots = y_n = 0$.

We construct the reduced OBDD of the transition relation of M_{2^n} as in Fig. 13.13 (c). The roots of the sub-OBDDs C and D have the following meanings:

C: leads to the 1-sink if and only if $|y_2 \ldots y_n| = |x_2 \ldots x_n| + 1$, $x_2 \ldots x_n \neq 11 \ldots 1$.

D: leads to the 1-sink if and only if $x_2 = \ldots = x_n = 1$, $y_2 = \ldots = y_n = 0$.

It can easily be checked that all the subfunctions represented in the newly introduced nodes are pairwise different. Therefore

$$size(M_{2^n}) = size(M_{2^{n-1}}) - 3 + 8 = size(M_{2^{n-1}}) + 5.$$

□

Hence, for each encoding of an autonomous counter there exists a re-encoding which leads to an OBDD of linear size. As according to Theorem 13.16 most counter encodings lead to OBDDs of exponential size, the gain between the original OBDD and the OBDD after performing a suitable re-encoding is exponential in most cases. Now the aim is to *find* those re-encodings which lead to small OBDD sizes.

Similarly to dynamic reordering or to linear sifting, when performing re-encodings, it is also promising to combine local operations which can be performed quickly. In particular, here, the set of elementary linear transformations introduced in Section 13.3 provides an attractive class, too. If for example each state is described by a bit-string $(q_1, \ldots, q_n)$, then the elementary linear transformation $q_i \mapsto q_i \oplus q_j$ defines the following re-encoding:

$$(q_1, \ldots, q_n) \mapsto (q_1, \ldots, q_{i-1}, q_i \oplus q_j, q_{i+1}, \ldots, q_n).$$

Example 13.20. Figure 13.14 shows the effect of an encoding transformation $q_1 \mapsto q_1 \oplus q_2$.   ◇

Original encoding		New encoding	
q_1	q_2	q_1^{new}	q_2^{new}
0	0	0	0
0	1	1	1
1	0	1	0
1	1	0	1

Figure 13.14. The elementary linear transformation $q_1 \mapsto q_1 \oplus q_2$ on the state space

13.5 References

The concept of transformed decision diagrams as well as the use of type-based cube transformations is due to Bern, Meinel, and Slobodová [BMS95]. The use of linear transformations for optimization was proposed by Meinel and Theobald [MT96], its realization through the linear sifting algorithm was done Meinel, Somenzi, and Theobald [MST97]. Finally, optimization of OBDDs by means of encoding transformations goes back to the papers [MT96, MT97].

Bibliography

[AHU74]   AHO, A. V., J. E. HOPCROFT and J. D. ULLMAN: *The Design and Analysis of Algorithms.* Addison-Wesley, Reading, MA, 1974.

[BBEL96]   BEER, I., S. BEN-DAVID, C. EISNER and A. LANDVER: *RuleBase: An industry-oriented formal verification tool.* In *Proc. 33rd ACM/IEEE Design Automation Conference (Las Vegas, NV)*, pages 655–660, 1996.

[BC95]   BRYANT, R. E. and Y.-A. CHEN: *Verification of arithmetic circuits with binary moment diagrams.* In *Proc. 32nd ACM/IEEE Design Automation Conference (San Francisco, CA)*, pages 535–541, 1995.

[BCL⁺94]   BURCH, J. R., E. M. CLARKE, D. E. LONG, K. L. MCMILLAN and D. L. DILL: *Symbolic model checking for sequential circuit verification.* IEEE Transactions on Computer-Aided Design of Integrated Circuits, 13:401–424, 1994.

[BFG⁺93]   BAHAR, R. I., E. A. FROHM, C. M. GAONA, G. D. HACHTEL, E. MACII, A. PARDO and F. SOMENZI: *Algebraic decision diagrams and their applications.* In *Proc. IEEE International Conference on Computer-Aided Design (Santa Clara, CA)*, pages 188–191, 1993.

[BGMS94]   BERN, J., J. GERGOV, CH. MEINEL and A. SLOBODOVÁ: *Boolean manipulation with free BDDs. First experimental results.* In *Proc. European Design Automation Conference*, pages 200–207, 1994.

[BHMS84]   BRAYTON, R. K., G. HACHTEL, C. MCMULLEN and A. SANGIOVANNI-VINCENTELLI: *Logic Minimization Algorithms for VLSI Synthesis.* Kluwer Academic Publishers, Boston, MA, 1984.

[BHS⁺96]   BRAYTON, R. K., G. D. HACHTEL, A. SANGIOVANNI-VINCENTELLI, F. SOMENZI and OTHERS: *VIS: A system for verification and synthesis.* In *Proc. Computer-Aided Verification '96*, volume 1102 of *Lecture Notes in Computer Science*, pages 428–432. Springer, Berlin, 1996.

[Big90]   BIGGS, N. L.: *Discrete Mathematics.* Clarendon Press, Oxford, 1990.

[BLPV95]   BORMANN, J., J. LOHSE, M. PAYER and G. VENZL: *Model checking in industrial hardware design.* In *Proc. 32nd ACM/IEEE Design Automation Conference (San Francisco, CA)*, pages 298–303, 1995.

[Blu84]   BLUM, N.: *A Boolean function requiring 3n network size.* Theoretical Computer Science, 28:337–345, 1984.

[BMS95]   BERN, J., CH. MEINEL and A. SLOBODOVÁ: *OBDD-Based Boolean manipulation in CAD beyond current limits.* In *Proc. 32nd ACM/IEEE Design Automation Conference (San Francisco, CA)*, pages 408–413, 1995.

[Boo54]   BOOLE, G.: *An Investigation of the Laws of Thought.* Walton, London, 1854. Reprinted by Dover Books, New York, 1954.

[BRB90]   BRACE, K. S., R. L. RUDELL and R. E. BRYANT: *Efficient implemen-tation of a BDD package.* In *Proc. 27th ACM/IEEE Design Automation Conference (Orlando, FL)*, pages 40–45, 1990.

[Bro90]   BROWN, F. M.: *Boolean Reasoning.* Kluwer Academic Publishers, Boston, MA, 1990.

[BRSW87] BRAYTON, R. K., R. RUDELL, A. SANGIOVANNI-VINCENTELLI and A. R. WANG: *MIS: A multiple-level interactive logic optimization sys-tem.* IEEE Transactions on Computer-Aided Design of Integrated Cir-cuits and Systems, 6:1062–1081, 1987.

[Bry86]   BRYANT, R. E.: *Graph-based algorithms for Boolean function manipu-lation.* IEEE Transactions on Computers, C–35:677–691, 1986.

[Bry91]   BRYANT, R. E.: *On the complexity of VLSI implementations and graph representations of Boolean functions with application to integer multi-plication.* IEEE Transactions on Computers, C–40:205–213, 1991.

[Bry92]   BRYANT, R. E.: *Symbolic Boolean manipulation with ordered binary decision diagrams.* ACM Computing Surveys, 24(3):293–318, 1992.

[BW96]   BOLLIG, B. and I. WEGENER: *Improving the variable ordering of OBDDs is NP-complete.* IEEE Transactions on Computers, 45:993–1002, 1996.

[CBM89]   COUDERT, O., C. BERTHET and J. C. MADRE: *Verification of syn-chronous sequential machines using symbolic execution.* In *Proc. Work-shop on Automatic Verification Methods for Finite State Machines*, volume 407 of *Lecture Notes in Computer Science*, pages 365–373. Springer, Berlin, 1989.

[CES86]   CLARKE, E. M., E. A. EMERSON and A. P. SISTLA: *Automatic ver-ification of finite-state concurrent systems using temporal logic specifi-cations.* ACM Transactions on Programming Languages and Systems, 8:244–263, 1986.

[CLR90]   CORMEN, T. H., C. E. LEISERSON and R. L. RIVEST: *Introduction to Algorithms.* MIT Press, Cambridge, MA, 1990.

[CM95]   COUDERT, O. and J. C. MADRE: *The implicit set paradigm: A new approach to finite state system verification.* Formal Methods in System Design, 6(2):133–145, 1995.

[CMZ+93] CLARKE, E. M., K. L. MCMILLAN, X. ZHAO, M. FUJITA and J. C.-Y. YANG: *Spectral transforms for large Boolean functions with application to technology mapping.* In *Proc. 30th ACM/IEEE Design Automation Conference (Dallas, TX)*, pages 54–60, 1993.

[Cob66]   COBHAM, A.: *The recognition problem for the set of perfect squares.* In *7th SWAT*, 1966.

[Cou94]   COUDERT, O.: *Two-level logic minimization: An overview.* INTEGRA-TION, the VLSI journal, 17:97–140, 1994.

[DST+94] DRECHSLER, R., A. SARABI, M. THEOBALD, B. BECKER and M. A. PERKOWSKI: *Efficient representation and manipulation of switching functions based on ordered Kronecker functional decision diagrams.* In *Proc. 31st ACM/IEEE Design Automation Conference (San Diego, CA)*, pages 415–419, 1994.

[DTB96]   DRECHSLER, R., M. THEOBALD and B. BECKER: *Fast OFDD-based minimization of fixed polarity Reed-Muller expressions.* IEEE Transac-tions on Computers, C–45:1294–1299, 1996.

[Ehr10]   EHRENFEST, P.: *Review of L. Couturat, 'The Algebra of Logic'*. Journal
            Russian Physical & Chemical Society, 42:382, 1910.

[FHS78]   FORTUNE, S., J. HOPCROFT and E. M. SCHMIDT: *The complexity of
            equivalence and containment for free single variable program schemes.*
            In *Proc. International Colloquium on Automata, Languages and Pro-
            gramming*, volume 62 of *Lecture Notes in Computer Science*, pages
            227–240. Springer, Berlin, 1978.

[FMK91]   FUJITA, M., Y. MATSUNAGA and T. KAKUDA: *On variable ordering of
            binary decision diagrams for the application of multi-level logic synthe-
            sis.* In *Proc. European Design Automation Conference (Amsterdam)*,
            pages 50–54, 1991.

[FS90]   FRIEDMAN, S. J. and K. J. SUPOWIT: *Finding the optimal variable or-
            dering for binary decision diagrams.* IEEE Transactions on Computers,
            39:710–713, 1990.

[GB94]   GEIST, D. and I. BEER: *Efficient model checking by automated ordering
            of transition relation partitions.* In *Proc. Computer-Aided Verification*,
            volume 818, pages 299–310, 1994.

[GJ78]   GAREY, M. R. and M. JOHNSON: *Computers and Intractibility: A
            Guide to the Theory of NP-Completeness.* W. H. Freeman, San Fran-
            cisco, CA, 1978.

[GM94a]   GERGOV, J. and CH. MEINEL: *Efficient analysis and manipulation of
            OBDDs can be extended to FBDDs.* IEEE Transactions on Computers,
            43(10):1197–1209, 1994.

[GM94b]   GERGOV, J. and CH. MEINEL: *On the complexity of analysis and ma-
            nipulation of Boolean functions in terms of decision diagrams.* Infor-
            mation Processing Letters, 50:317–322, 1994.

[HCO74]   HONG, S., R. CAIN and D. OSTAPKO: *MINI: A heuristic approach
            for logic minimization.* IBM Journal of Research and Development,
            18:443–458, 1974.

[HLJ$^+$89] HACHTEL, G., M. LIGHTNER, R. JACOBY, C. MORRISON, P. MOCEYU-
            NAS and D. BOSTICK: *BOLD: The Boulder Optimal Logic Design Sys-
            tem.* In *Hawaii International Conference on System Sciences*, 1989.

[HS96]   HACHTEL, G. and F. SOMENZI: *Logic Synthesis and Verification Algo-
            rithms.* Kluwer Academic Publishers, Boston, MA, 1996.

[HU78]   HOPCROFT, J. E. and J. D. ULLMAN: *Introduction to Automata The-
            ory, Languages, and Computation.* Addison-Wesley, Reading, MA,
            1978.

[ISY91]   ISHIURA, N., H. SAWADA and S. YAJIMA: *Minimization of binary deci-
            sion diagrams based on the exchanges of variables.* In *Proc. IEEE In-
            ternational Conference on Computer-Aided Design (Santa Clara, CA)*,
            pages 472–475, 1991.

[Kar88]   KARPLUS, K.: *Representing Boolean functions with If-Then-Else
            DAGs.* Technical Report UCSC-CRL-88-28, Computer Engineering,
            University of California at Santa Cruz, 1988.

[KSR92]   KEBSCHULL, U., E. SCHUBERT and W. ROSENSTIEL: *Multilevel logic
            synthesis based on functional decision diagrams.* In *Proc. European
            Design Automation Conference*, pages 43–47, 1992.

[Lee59]   LEE, C. Y.: *Representation of switching circuits by binary decision
            programs.* Bell System Technical Journal, 38:985–999, 1959.

[Lon93]   LONG, D.: *Model Checking, Abstraction and Compositional Verification*. PhD thesis, Carnegie Mellon University, 1993.

[LPV94]   LAI, Y.-T., M. PEDRAM and S. B. K. VRUDHULA: *EVBDD-based algorithms for integer linear programming, spectral transformation and function decomposition*. IEEE Transactions on Computer-Aided Design of Integrated Circuits and Systems, 13:959–975, 1994.

[Mas76]   MASEK, W.: *A fast algorithm for the string editing problem and decision graph complexity*. Master's thesis, MIT, 1976.

[MB88]   MADRE, J.-C. and J.-P. BILLON: *Proving circuit correctness using formal comparison between expected and extracted behaviour*. In *Proc. 25th ACM/IEEE Design Automation Conference (Anaheim, CA)*, pages 205–210, 1988.

[McM93]   McMILLAN, K. L.: *Symbolic Model Checking*. Kluwer Academic Publishers, Boston, MA, 1993.

[Mei89]   MEINEL, CH.: *Modified Branching Programs and Their Computational Power*, volume 370 of *Lecture Notes in Computer Science*. Springer, Berlin, 1989. Reprinted by World Publishing Corporation, Beijing, 1991.

[Mic94]   MICHELI, G. DE: *Synthesis and Optimization of Digital Circuits*. McGraw-Hill, New York, NY, 1994.

[Min93]   MINATO, S.: *Zero-suppressed BDDs for set manipulation in combinatorial problems*. In *Proc. 30th ACM/IEEE Design Automation Conference (Dallas, TX)*, pages 272–277, 1993.

[Min96]   MINATO, S.: *Binary Decision Diagrams and Applications for VLSI CAD*. Kluwer Academic Publishers, Boston, MA, 1996.

[MIY90]   MINATO, S., N. ISHIURA and S. YAJIMA: *Shared binary decision diagrams with attributed edges*. In *Proc. 27th ACM/IEEE Design Automation Conference (Florida, FL)*, pages 52–57, 1990.

[Mor92]   MORET, B. M. E.: *Decision trees and diagrams*. ACM Computing Surveys, 14(4):593–623, 1992.

[MS94]   MEINEL, CH. and A. SLOBODOVÁ: *On the complexity of constructing optimal ordered binary decision diagrams*. In *Proc. Mathematical Foundations in Computer Science*, volume 841 of *Lecture Notes in Computer Science*, pages 515–524, 1994.

[MS97]   MEINEL, CH. and A. SLOBODOVÁ: *Speeding up variable ordering of OBDDs*. In *Proc. International Conference on Computer Design (Austin, TX)*, 1997.

[MS98]   MEINEL, CH. and H. SACK: *⊕-OBDDs - a BDD structure for probabilistic verification*. In *International Workshop on Logic Synthesis (Lake Tahoe, CA)*, 1998.

[MST97]   MEINEL, CH., F. SOMENZI and T. THEOBALD: *Linear sifting of decision diagrams*. In *Proc. 34th ACM/IEEE Design Automation Conference (Anaheim, CA)*, pages 202–207, 1997.

[MT96]   MEINEL, CH. and T. THEOBALD: *Local encoding transformations for optimizing OBDD-representations of finite state machines*. In *Proc. International Conference on Formal Methods in Computer-Aided Design (Palo Alto, CA)*, volume 1166 of *Lecture Notes in Computer Science*, pages 404–418. Springer, Berlin, 1996.

[MT97]   MEINEL, CH. and T. THEOBALD: *On the influence of the state encoding on OBDD-representations of finite state machines*. In *Mathematical*

Foundations of Computer Science (Bratislava), volume 1295 of *Lecture Notes in Computer Science*, pages 408–417. Springer, Berlin, 1997.

[MT98]   MEINEL, CH. and T. THEOBALD: *Ordered binary decision diagrams and their significance in computer-aided design of VLSI circuits.* Bulletin of the European Association for Theoretical Computer Science, (64):171–187, 1998.

[MWBS88]   MALIK, S., A. WANG, R. K. BRAYTON and A. SANGIOVANNI-VINCENTELLI: *Logic verification using binary decision diagrams in a logic synthesis environment.* In *Proc. IEEE International Conference on Computer-Aided Design (Santa Clara, CA)*, pages 6–9, 1988.

[Pap94]   PAPADIMITRIOU, C. H.: *Computational Complexity.* Addison-Wesley, Reading, MA, 1994.

[Pon95]   PONZIO, S.: *Restricted Branching Programs and Hardware Verification.* PhD thesis, MIT, 1995.

[PS95]   PANDA, S. and F. SOMENZI: *Who are the variables in your neighborhood.* In *Proc. IEEE International Conference on Computer-Aided Design (San José, CA)*, pages 74–77, 1995.

[Rud93]   RUDELL, R.: *Dynamic variable ordering for ordered binary decision diagrams.* In *Proc. IEEE International Conference on Computer-Aided Design (Santa Clara, CA)*, pages 42–47, 1993.

[Sha38]   SHANNON, C. E.: *A symbolic analysis of relay and switching circuits.* Transactions American Institute of Electrical Engineers, 57:713–723, 1938.

[Sha49]   SHANNON, C. E.: *The synthesis of two-terminal switching circuits.* Bell System Technical Journal, 28:59–98, 1949.

[Sie94]   SIELING, D.: *Algorithmen und untere Schranken für verallgemeinerte OBDDs.* PhD thesis, Universität Dortmund, 1994.

[Som96a]   SOMENZI, F.: *Binary Decision Diagrams.* Lecture Notes (University of Colorado, Boulder), 1996.

[Som96b]   SOMENZI, F.: *CUDD: Colorado University Decision Diagram Package.* ftp://vlsi.colorado.edu/pub/, 1996.

[SSM+92]   SENTOVICH, E. M., K. J. SINGH, C. MOON, H. SAVOJ, R. K. BRAYTON and A. SANGIOVANNI-VINCENTELLI: *Sequential circuit design using synthesis and optimization.* In *Proc. International Conference on Computer Design (Cambridge, MA)*, pages 328–333, 1992.

[SW93a]   SIELING, D. and I. WEGENER: *NC-algorithms for operations on binary decision diagrams.* Parallel Processing Letters, 3:3–12, 1993.

[SW93b]   SIELING, D. and I. WEGENER: *Reduction of OBDDs in linear time.* Information Processing Letters, 48:139–144, 1993.

[SW95a]   SCHRÖER, O. and I. WEGENER: *The theory of zero-suppressed BDDs and the number of knight's tours.* In *Proc. Workshop on Applications of the Reed-Muller Expansion*, pages 38–45, 1995.

[SW95b]   SIELING, D. and I. WEGENER: *Graph driven BDDs – a new data structure for Boolean functions.* Theoretical Computer Science, 141:283–310, 1995.

[SW97]   SAVICKÝ, D. and I. WEGENER: *Efficient algorithms for the transformation between different types of binary decision diagrams.* Acta Informatica, 34:245–256, 1997.

[THY93]   TANI, S., K. HAMAGUCHI and S. YAJIMA: *The complexity of the optimal variable ordering problems of shared binary decision diagrams.*

In *Proc. International Symposium on Algorithms and Computation*, volume 762 of *Lecture Notes in Computer Science*, pages 389–398. Springer, Berlin, 1993.

[TI94]   TANI, S. and H. IMAI: *A reordering operation for an ordered binary decision diagram and an extended framework for combinatorics of graphs*. In *Proc. International Symposium on Algorithms and Computation*, volume 834 of *Lecture Notes in Computer Science*, pages 575–592. Springer, Berlin, 1994.

[Weg87]   WEGENER, I.: *The Complexity of Boolean Functions*. John Wiley & Sons and Teubner-Verlag, New York-Stuttgart, 1987.

[Weg94]   WEGENER, I.: *Efficient data structures for Boolean functions*. Discrete Mathematics, 136:347–372, 1994.

Index

Printing: Mercedesdruck, Berlin
Binding: Buchbinderei Lüderitz & Bauer, Berlin